PRAISE FOR

the cheating culture

"This is a breathtaking book." —*Los Angeles Times Book Review*

"Well-constructed, civic-minded... full of compelling statistics and anecdotes... relentlessly big picture."
—*The New York Times Book Review*

"For as long as people have come under pressure to produce, to excel or to rise, some people have tried to use the shortcuts that cheating offers. But there is much reason to believe that these days, as David Callahan puts it in *The Cheating Culture*, 'cheating is everywhere.'... Watch what we do, not what we say or read; our real-world *Book of Virtues* is in serious need of rewriting."
—Jonathan Yardley, *Washington Post*

"Provocative... That Americans cheat more than they used to sounds like an impossible hypothesis to prove. And yet, Callahan's book is thick with convincing examples." —*Boston Globe*

"[A] damning and persuasive critique of America's new economic life." —*Esquire*

"There is no shortage of evidence of cynical crimes pervading the institutions of the United States... Dozens of books and hundreds of newspaper, magazine and journal articles have examined this phenomenon. New ones come out daily. None I have yet seen

does it with the anger, vigor and persuasiveness of *The Cheating Culture*."
—*Baltimore Sun*

"Callahan has done us a good turn by confronting the question of 'why do Americans do wrong?' Here's hoping *The Cheating Culture* gets past the sound bites and pundits and fuels our energetic introspection and real willingness to become better acquainted with honesty."
—*Seattle Times*

"What used to be confined to taxes and matrimony has branched out into every corner of society... Using a mirror to show us ourselves, Callahan has explained powerfully that we need honor now, right now."
—*Houston Chronicle*

"In what he aptly describes as the Cheating Culture, David Callahan identifies the rising tide of cheating across American society as the inevitable moral downside of unfettered market economics. This book has an important role to play in the fight against the culture of avarice." —Kevin Phillips, author of *American Dynasty*

"Here, finally, a lucid explanation for why America seems on its way to becoming a nation of cheaters. Cheating begets more cheating, from our boardrooms to our classrooms. With verve and insight, Callahan shows that too many Americans feel the dice are already loaded, so they might as well give them an extra roll when no one's looking. His solutions are as provocative as is his diagnosis. Buy this book (don't steal it) and take to heart its wisdom."
—Robert B. Reich, Author of *The Future of Success*
and former U.S. Secretary of Labor

"Presents a Technicolor panorama of shameful behavior."
—*The Village Voice*

"One of the most provocative books of the season." —*Fast Company*

"The author provides persuasive evidence that our society is riddled with dishonesty, people who are willing to 'cheat' to get a buck or who consider what they do so prevalent—'Everyone's doing it'—that they do not even analyze its morality."

—*Deseret Morning News*

"Callahan makes his case by piling on story after story of selfishness exercised from boardrooms to classrooms... In the end, *The Cheating Culture* persuades us of the permanence, as well as the gravity, of this problem." —*BookPage*

"A hundred years ago, the progressive reformer could draw on a coherent worldview shaped by Social Gospel Protestantism. For better and worse, our culture lacks that coherence today. But if we ever do try to overcome our divisions and regenerate our public life, we will find few more thoughtful guides than David Callahan."

—*In These Times*

"Meticulously researched." —*Booklist*

"The book's strength lies in tying together assorted detailed descriptions of cheating throughout the system and explaining the connections between disparate acts like resume inflation, tax evasion and illegal downloads. [Callahan] offers straightforward, commonsensical solutions." —*Publishers Weekly*

the cheating culture

DAVID CALLAHAN

the cheating culture

Why More Americans Are Doing Wrong to Get Ahead

A HARVEST BOOK • HARCOURT, INC.
ORLANDO AUSTIN NEW YORK SAN DIEGO TORONTO LONDON

www.HarcourtBooks.com

Library of Congress Cataloging-in-Publication Data
Callahan, David, 1965–
The cheating culture: why more Americans are doing wrong to get ahead/
David Callahan.
p. cm.
Includes bibliographical references and index.
ISBN 0-15-101018-8
ISBN 0-15-603005-5 (pbk.)
1. Business ethics. 2. Professional ethics. 3. Social ethics.
4. United States—Moral conditions. I. Title.
HF5387.C334 2004
174—dc22 2003015529

Text set in Adobe Jenson
Designed by Cathy Riggs

Printed in the United States of America

First Harvest edition 2004
A C E G I K J H F D B

contents

preface

My friend Peter went through a shoplifting phase a few years back. His thefts reflected his tastes—$40 bottles of Bordeaux, for example—and though Peter showed no signs of kleptomania, his friends agreed that he must have mental problems. Why else would a normal adult be shoplifting?

Last April, my friend Max filed taxes in Connecticut, where he used to live, even though he now resides in New York City. Dodging the city's killer taxes saved him about $3,000. Did anyone think Max was disturbed? Hardly. His friends thought he was smart to keep his legal address in Connecticut.

Petty shoplifting is a misdemeanor. Tax evasion is a felony that can yield prison time. If anyone has a screw loose, it's really Max, not Peter. Right? Well, yes and no. And this funny dichotomy says a lot about morality in America these days, as I've learned while writing this book.

In the fall of 2001, when Enron collapsed into a heap of debts and lawsuits, I was working on a history of the Harvard Business School class of 1949. Most of its members are now in their late seventies, and many ran big companies in their day. When I asked them about the corporate scandals, they'd shake their heads in disgust—disgust at the bloated pay packages, the gilded perks, and most of all at the pervasive lying by CEOs. It's about values, the '49ers said, in trying to explain what went wrong in the executive suites of Enron, Tyco, and so many other companies. Today's business values normalize felonious behavior; yesterday's values were less tolerant of such behavior. To the '49ers, the corporate scandals were almost that simple.

This was not reassuring to hear. It meant that, sure, we could pass stiffer laws and get new watchdogs. We could toss CEOs in prison and throw away the key. But such a crackdown would only get America so far, as long as business leaders were greedy, self-centered, and materialistic.

As I mulled over these questions of values, I began noticing other stories in the news about cheating. The historian Stephen Ambrose was enmeshed in a plagiarism scandal; Princeton's admissions office was in trouble for breaking into Yale's computer; Alabama's football team was put on probation for recruitment violations; the IRS reported that tax evasion was up sharply. There was even a story about cheating on the hallowed American ground of state fairs, as contestants misrepresented the real weight of livestock to win cash prizes. I wondered, what's going on here? Why so much cheating? Is there more cheating now than in the past, as it seems? And is it all connected?

To answer these questions, I set out to explore what changes in American life might be leading us to cheat more. I cast a wide net in my research, looking at different professions, our government and legal system, the economy and culture, and people's values. I've drawn on government reports and statistics, studies by social scientists, public opinion polls, histories of different professions, and a mountain of journalistic investigations of scandals and cheating incidents. My research assistants and I also conducted over eighty interviews with people involved in the cheating culture in one way or another: parents, students, teachers, coaches, athletes, experts in business ethics, stock analysts, lawyers, accountants, doctors, and law enforcement officials. These interviews were immensely helpful, especially those where cheaters talked openly about their motives. For obvious reasons, many of the people interviewed did not want their names used.

This is a dark book in some ways. An increase in cheating reflects deep anxiety and insecurity in America nowadays, desperation even, as well as arrogance among the rich and cynicism among ordinary people. Many of the stories that follow are very troubling; some are tragic. And yet there is real hope here, too. Much cheating, as we'll see, can be traced to conditions that we have the power to change—from how much security our economy affords people, to how well government polices business, to the ethical climate in our schools. We can make different choices in the years ahead, and I suggest a number of such choices in my concluding chapter.

I owe a large debt of gratitude to those who helped make this book possible. In the back of the book, I've listed all the people who shared their expertise and experiences. Here, I'd like to thank those who worked closest with me in developing the book: Andrea Schulz, my editor at Harcourt, who brought extraordinary talent to the task of improving successive drafts of the manuscript; Andrew Stuart, my literary agent; LeeAnna Keith and Carolyn Rendell, my research assistants; and my colleagues at Demos, especially Miles Rapoport, who has been unfailingly supportive of this project. Also, a special thanks to Wendy Paris for her encouragement and thoughtful comments on the manuscript—and for her patience and love.

David Callahan
New York
July 2003

the cheating culture

CHAPTER ONE

"Everybody Does It"

I PLAYED A LOT OF MONOPOLY GROWING UP. LIKE MOST PLAYERS of the game, I loved drawing a yellow Community Chest card and discovering a "bank error" in my favor—"Collect $200!" It never occurred to me not to take the cash. After all, banks have plenty of money and if one makes an error in your favor, why argue?

I haven't played Monopoly in twenty years, but I'd still take the $200 today. And what if a real bank made an error in my favor? That would be a tougher dilemma.

Such things do happen.

Just to the east of where the Twin Towers once stood is a twenty-six-story office building that houses the Municipal Credit Union of New York City. The credit union has 300,000 members—federal, state, and city government employees—and over $1 billion in assets. Although a number of buildings near Ground Zero sustained serious damage when the towers came down, the MCU's glass-and-steel building on Cortlandt Street survived unscathed. However, the credit union did suffer a major computer failure that severed its link to the New York Cash Exchange (NYCE), the largest network of automatic teller machines in the Northeast.

The network managers at NYCE quickly detected the severed link. The problem meant that while credit union members could withdraw money at cash machines, NYCE couldn't immediately track these transactions or prevent members from overdrawing their accounts. NYCE leaders managed to get through to the credit union staff, even though the organization was in chaos. They posed the following choice: With just a few strokes on a computer keyboard, NYCE could cut off all cash withdrawals until the severed link was restored—which could take several weeks—or NYCE could let the cash keep flowing and sort out the withdrawal records later. Theoretically, anyone with a credit union ATM card could take out as much money as they wanted. The credit union would have to assume that risk. What did it want to do?

The Municipal Credit Union of New York is one of the oldest credit unions in America, founded more than eight decades ago. It is guided by an ethos of self-help and pooled aspirations. Many of its members are firemen and policemen and, in the wake of the attacks, it was widely assumed that some of these people had perished just across the street from the MCU's office. There was no way the credit union would prevent its members and their families from accessing their money at a time of crisis. Thomas Siciliano, the general counsel of the credit union, said later: "We felt it would have hurt them badly and added to the chaos of the city." The MCU trusted them to use their ATM cards responsibly.

Credit union members realized early on that their ATM use wasn't monitored and that there was no limit to how much cash they could take out. As word spread, withdrawals skyrocketed. As many as 4,000 members overdrew their accounts, some by as much as $10,000. One member used his card more than 150 times between late September and mid-October.

In November, the computer link with NYCE was finally restored. As the credit union got back to normal, it pieced together the full record of cash withdrawals after September 11. Those who had overdrawn their accounts had left a substantial electronic trail, and the MCU set about tracking them down. Siciliano led this work. He initially suspected that most of the members' overdrawing had occurred by accident, or maybe was prompted by emergency needs. The MCU assumed the best of its members, even those with average bank balances of less than $100 who had withdrawn thousands of dollars in just a few weeks. "We try to understand people," Siciliano says. "We're not just about the bottom line."

The MCU sent letters to those with overdrawn accounts listing the money that was missing and asking for repayment. While some money was repaid, many letters got no response. More letters were sent—notarized letters with threats. After months of appeals, $15 million was still missing. At that point, the MCU called in the authorities. A criminal investigation, led by Manhattan District Attorney Robert Morgenthau and the New York City Police Department, extended into the following summer. It resulted in scores of arrests.

A FEW BLOCKS AWAY from the credit union's offices, another investigation was reaching its climax in the spring of 2002, this one at Merrill Lynch's newly repaired global headquarters on Vesey Street. After September 11, Merrill Lynch had scattered 9,000 employees around back-office facilities in New Jersey and midtown. Months passed before it was able to move back downtown. When Merrill did return, morale at the company was low. Huge layoffs had depleted its ranks and profits were down in the new bear market. Worse, Merrill found itself cornered in a criminal probe led by New York State Attorney General Eliot Spitzer.

Before his assault on Wall Street made him famous, Spitzer was an obscure state official. Those who did know him were reminded of a character straight out of early-twentieth-century

America. Wealthy by birth, with a father who bankrolled his political career, Spitzer is a muckraking crusader for the public interest.

Merrill Lynch had come to Spitzer's attention in a circuitous fashion. In early 2001, a Queens pediatrician named Debases Kanjilal hired a lawyer to pursue a civil suit against Merrill. Kanjilal was among the legions of investors who got burned when the NASDAQ cratered in 2000. Specifically, he had lost $500,000 on a single Internet stock, InfoSpace. Kanjilal's instinct had been to sell InfoSpace when it was trading at $60 a share. But his broker at Merrill Lynch had urged him to hold on to the stock, advice that reflected Merrill's public research reports that recommended InfoSpace as a "buy" stock. Standing behind those research reports, and affirming their recommendations in his TV appearances, was Merrill's star analyst and "Internet stock guru," Henry Blodget.

It is hard today to appreciate the influence once wielded by Blodget. Just over thirty years old in 2000, Blodget was a Yale grad who had never aspired to stardom on Wall Street. He had tried instead to make it as a writer, and when that didn't work out, his father rescued him from unemployment by helping him land a position at Prudential Securities. Blodget's career was unremarkable until he shot to fame in 1998 with his prediction that Amazon's stock would reach the unthinkable price of $400 a share. When the stock did, in fact, hit that level a month later, Blodget was hailed as an oracle. Shortly thereafter he moved to Merrill Lynch with a $3 million contract. There, he reigned as the single most visible adviser to investors hoping to score big in the Internet gold rush. Blond and affable, with telegenic good looks, Blodget was everywhere with his stock predictions as well as broader prognostications about the new economy.

What Blodget didn't mention to CNBC junkies or Merrill Lynch's own clients was that his role at Merrill went far beyond analyzing stocks. Like other star analysts of the time, he also became deeply involved in Merrill's investment banking business, helping to bring Internet companies—and fat underwriting fees—to Merrill. One of the companies Merrill's investment banking division represented was Go2Net, a company that InfoSpace was in the process of purchasing in 2000. Merrill had a financial interest in InfoSpace's stock price staying high so that the deal would go through.

Debases Kanjilal held on to his InfoSpace stock even as it declined steadily. Finally he sold at $11 a share and took a staggering loss. At the time Kanjilal sold, Merrill and Blodget were continuing to recommend InfoSpace to investors. Kanjilal's losses were part of an estimated $4 trillion that investors lost when NASDAQ crashed. Big-name analysts hyped many sinking tech stocks with the same enthusiasm they'd shown in pumping them up. For example, as of May 2001, Morgan Stanley's top Internet analyst, Mary Meeker, was still bestowing her once-coveted "outperform" rating on Priceline, then down from $162 to $4, and on Yahoo!, down from $237 to $19.50.

Kanjilal's lawsuit against Merrill Lynch attracted the attention of Eliot Spitzer's office not long after it was filed. Initiating a criminal investigation, Spitzer uncovered a shocking pattern of public deceit and conflict of interest at Merrill Lynch. He found e-mails by Henry Blodget privately ridiculing the same stocks that he and Merrill were publicly pushing. "A piece of junk," Blodget had called InfoSpace, even as he recommended it. He privately called other stocks a "pos," or piece of shit. Spitzer also found a memo in which Blodget detailed the compensation he deserved for bringing in investment banking business—a memo that flatly contradicted Merrill's claims that analysts were not rewarded for playing such a role. As a result of the investigation, Spitzer charged that Merrill Lynch's "supposedly independent and objective investment advice was tainted and biased by the desire to aid Merrill Lynch's investment banking business." In Spitzer's view, the behavior by Merrill and Blodget constituted securities fraud, a serious felony.[1]

Spitzer's evidence against Merrill Lynch resulted in the company agreeing to pay a $100 million settlement. This case turned out to be just the first step in a larger investigation of other top Wall Street firms that had engaged in a range of abuses by insiders, which culminated in a historic $1.4 billion settlement in 2003.

And what happened to Blodget? Not much. Saying he wanted a "lifestyle change," Blodget had accepted a November 2001 buyout offer from Merrill worth an estimated $5 million. He spent his days working on a book for Random House and meeting regularly with lawyers. In 2003, Blodget settled with Spitzer's office, agreeing to pay a $4 million penalty—yet admitting no wrongdoing. The settlement was easy enough to afford. Blodget had pulled in nearly $20 million during his brief star turn at Merrill.

HENRY BLODGET and the ATM looters have nothing in common and much in common. Blodget was among the ranks of the big winners in the new economy—the very top earners who saw unprecedented income gains during the boom of the 1990s. His education and background had helped him to secure his place in the Winning Class: successful parents, private schools, Yale University, connections on Wall Street.

The ATM looters, by contrast, were among the far larger ranks of Americans who had either stayed put economically or realized only modest gains during the boom years. They occupied the lower rungs of what Robert Reich has called the Anxious Class, and the 1990s were not easy for them. Although median wages for workers near the bottom crept up in the latter part of the decade, these gains did not make up for wage losses since the late 1970s and, in any case, were wiped out by large increases in the cost of living across the New York area. Records from the DA's office indicated that most of the ATM looters lived paycheck to paycheck with little money in the bank for emergencies. Some had average balances below $100 for months on end.

Economically and culturally, Henry Blodget and the ATM looters might as well have lived on separate planets. What they have in common is that both did wrong—yet both squarely identified themselves as upstanding citizens. Many members of the Municipal Credit Union work for the very authorities that enforce law and order in New York City. They would never have contemplated robbing a bank. But, hey, if a cash machine starts spitting out free money, what are you going to do? Meanwhile, Blodget did not begin his career on Wall Street imagining that one day he'd end up in the crosshairs of the state attorney general, and in many ways Blodget was simply the fall guy at Merrill Lynch. A close reading of the e-mails uncovered by Spitzer shows that Blodget often caved to company pressures to hype stocks and was uncomfortable with his role. At Merrill, like many other financial services companies, the investment bankers were notorious for leaning heavily on the analysts to say the right things about the stocks of important clients.

Blodget made millions playing by the rules of a deeply corrupt game. Plenty of other analysts did the same thing and many thought nothing of it. "The system was sordid," says one analyst who worked at Prudential during this period. "But because everyone knew it was sordid, it no longer seemed sordid anymore." As the analysts saw it, the big institutional investors on Wall Street were not naïve, and they weren't stupid enough to believe even half of what research analysts tied to investment banks said about the companies their banks represented. "Everyone knew how the game was played," says the former Prudential analyst. Analysts hyped stocks because they had to, and serious investors simply ignored them.

The problem was that a star analyst like Blodget wasn't talking to insiders; he was on television, speaking to the public, and his recommendations were also heeded by Merrill Lynch brokers nationwide as they counseled clients on where to invest. "We are losing people money and I don't like it," one of Blodget's colleagues complained to him in an e-mail. "John and Mary Smith are losing their retirement."[2]

Blodget made an attractive poster boy for Wall Street corruption, just as he had been the perfect pitchman for the high-tech bull market. Yet ultimately there was nothing uniquely immoral about Henry Blodget. He found himself in a cheating culture and he went along.

It's easy to bash Blodget for getting rich in a corrupt system or the ATM looters for ripping off their own credit union. But these days many of us aren't behaving much better. In one area of American life after another—sports, business, law, education, science, medicine—more people seem to be cutting corners. Consider the following:

- A psychiatrist in Westchester County, Dr. Dana Luck, suddenly finds herself busy evaluating local teenagers for signs of even the slightest learning disability. She knows what is causing the spike in her business: a College Board ruling that students with disabilities who receive extra time on the SATs will no longer be identified to admissions officers as disabled. The wealthy parents coming to Dr. Luck want only one thing: an official diagnosis of disability that will allow their kids more time on the SATs. Dr. Luck finds nothing wrong with most of her young patients. But parents who keep "diagnosis shopping" can find a more compliant disability expert and, for the right price, get what they want. Meanwhile many poorer kids with learning disabilities go without the diagnoses they deserve because they can't afford the cost. Although it is well known that academic cheating by students has reached an all-time high, it's also true that parents and tutors and other adults are increasingly helping students do whatever it takes to get an edge in their high-stakes education careers. Money lubricates much of this corner cutting.

- A researcher at Harvard Medical School, David Franklin, takes a job as a "medical liaison" for a large pharmaceutical company. His job is to reach out to doctors and explain to them the many reasons why they should prescribe the company's new drug, Neurontin. Federal law says that drugmakers can only promote a drug for FDA-approved purposes. But Franklin is pressured by his superiors to promote Neurontin for a wide range of "off-label" uses, many of which are wholly untested and possibly dangerous for patients. He lies to doctors nearly every day, telling them anything that will get them excited about Neurontin. According to court records, his company also offers doctors large cash payments to push Neurontin's off-label uses to other doctors and to sign journal articles they didn't write touting the virtues of the drug, which haven't been verified by clinical trials. Do the doctors object? Hardly. Thousands of doctors pocket kickbacks to become Neurontin pushers. The Neurontin scandal is only one of many prescription drug scandals that have recently rocked the medical world.

- A reporter is writing an article on Ronald Zarrella, the CEO of Bausch & Lomb. Checking Zarrella's background, the reporter discovers that NYU's School of Business Administration has no record of the M.B.A. that Zarrella says he earned there. Confronted with this information, Zarrella confirms that, in fact, he did not get an M.B.A. from NYU as he had long contended. This revelation is just the latest in a spate of résumé-faking cases, including ones involving high-profile people like George O'Leary, the former Notre Dame football coach; Kenneth Lonchar, the former chief financial officer of Veritas software; and Sandra Baldwin, the former president of the U.S. Olympic Committee. Executive recruitment and employment agencies say the problem of misrepresentation by job seekers at every level has soared over the past decade and that up to half of résumés include lies.

- I'm out with a group of friends at dinner. The check comes to $141 and we split it evenly. Then one of my friends, a freelance writer, reaches for the receipt. "Anyone mind if I keep the receipt?" She's not asking whether there are any ethical objections to her writing off our expenditures for her taxes; she's wondering whether anyone else had hoped to do the same thing. Nobody objects on the latter grounds, and certainly not on the former. I'm not surprised. The IRS reports that tax evasion has gotten worse in recent years, costing the U.S. Treasury a minimum of $250 billion a year, and maybe twice that. Wealthy Americans are the biggest offenders, but sophisticated tax evasion is becoming a more populist activity. For example, as many as two million Americans now have illegal offshore bank accounts that they use to evade taxes, a problem that increased dramatically in the 1990s. Good weather, it turns out, is only part of the Caribbean's appeal.

- A leading high school basketball player named LeBron James, the next Michael Jordan some say, shows up one day at his school in Akron driving a new $50,000 Hummer H2 sports utility vehicle crammed with three TVs. The Ohio High School Athletic Association immediately launches an investigation, suspecting that the Hummer is a gift from a sports agent or university recruiter. James denies everything. My mom gave it to me, he says. Few believe that James's middle-class mother can afford a top-of-the-line Hummer, but no one can prove a violation of state rules. It's a typical episode in the money-saturated world of collegiate and professional sports, where recruiting violations, drug use, and other kinds of cheating—like Sammy Sosa's corked bat—are pervasive.

- A new technology is developed that allows ordinary Americans to engage in the large-scale theft of copyrighted materials. Use of the technology spreads rapidly, especially on college campuses, and results in hundreds of millions of dollars in lost sales revenue by companies and individual artists. This epidemic of stealing via Napster and other music file-sharing programs is accompanied by almost zero ethical second-guessing by users. The music industry, after all, is reviled for its greed and commercialism. Music piracy is nothing compared to the widespread theft of cable and satellite services. Americans are now stealing nearly $6 billion a year worth of paid television. Hooking up the neighbors so they can watch the *Sopranos*, too—*sans* a tribute to HBO—is considered the community-minded thing to do. Americans may be bowling alone, as Robert Putnam says, but increasingly we are stealing together.

- A former *New York Times* reporter decides to write a book about his stint at the newspaper. He's a young man who only spent a few years at the *Times* and he didn't win any Pulitzer Prizes. Still, the book proposal generates buzz and results in a mid-six-figure advance. The reporter, who previously could barely afford to pay his rent, is slated to become quite wealthy before the age of thirty. All in all, not a bad payoff for Jayson Blair, who fabricated quotes and other information in numerous stories. Blair's financial rewards easily outdo those of Stephen Glass, the disgraced *New Republic* writer who won a large advance from Simon and Schuster to write a novel about his dishonest career and was invited on *60 Minutes* to promote it. We're all used to the idea of tax evasion or cheating on Wall Street. But cheating by writers? Yes, indeed. An unprecedented number of cases of plagiarism and fraud have rocked the worlds of book publishing and journalism in recent years, including those involving historians Stephen Ambrose, Michael Bellesiles, and Doris Kearns Goodwin, and journalists Patricia Smith, Mike Barnicle, Michael Finkel, and others.

- A management consultant is out golfing with two CEOs who are negotiating a deal worth millions. He is shocked when the CEOs decide to bet an aspect of the deal worth $150,000 on the outcome of the golf game (company money, mind you). He is even more shocked when he sees one of the CEOs kick his opponent's ball into the woods to help him gain a winning advantage. In fact, none of this should come as a surprise. A 2002 survey of high-ranking corporate executives revealed that 82 percent admitted to cheating on the golf course. Why? Because playing golf is now a big part of networking and doing business in corporate America, and it doesn't look good to be a terrible player.

These stories are not isolated instances. They are part of a pattern of widespread cheating throughout U.S. society. By its nature cheating is intended to go undetected, and trends in unethical behavior are hard to document. Still, available evidence strongly suggests that Americans are not only cheating more in many areas but are also feeling less guilty about it. When "everybody does it," or imagines that everybody does it, a cheating culture has emerged.

Yet why all the cheating, and why now?

One might think that there'd be no shortage of possible explanations floating around for this crisis. America has been a nation of moralizers since the days of Benjamin Franklin, who advised in his 13 Virtues to "Imitate Jesus and Socrates"—a pretty high bar. But rarely has that cultural leaning been more pronounced than in recent decades. We have been living in the age of the Moral Majority and the Christian Coalition, the age of family values and zero tolerance. Religious figures and intellectuals and newspaper columnists have talked endlessly in recent years about moral issues large and small: teen pregnancy, school uniforms, violent video games, graffiti, pedophilia, welfare dependency, crime, drug use, and so forth. God, who previously didn't play much of a role in American politics, has come to be as omnipresent in election campaigns as corporate donors seeking favors.

Yet America's watchdogs of virtue have been largely silent about the new epidemic of cheating. To be sure, rampant cheating by students has begun to receive attention in the past several years. And the recent corporate scandals induced a media feeding frenzy. There have also been big stories about cheating by athletes, or tax evasion, or plagiarism by journalists. Still, there's been very little effort to connect all these dots and see them for what they represent: a profound moral crisis that reflects deep economic and social problems in American society.

Concerns about cheating do not jibe easily with the way that Americans have talked about values and personal responsibility since the early 1980s. That conversation has been orchestrated by conservatives and the religious right, while liberals—often uncomfortable talking about values—have largely kept their mouths shut. America's moral ills were defined in the '80s and '90s in terms that reflected traditional conservative worries, with a focus on things like crime, drugs, premarital sex, and divorce. Other concerns—little problems like greed, envy, materialism, and inequality—have been excluded from the values debate.

But lately conservatives haven't had much to complain about. Many aspects of Americans' personal behavior have changed in recent years. Crime is down. Teenage pregnancy is down. Drunk driving is down. Abortion is down. The use of tobacco and illicit drugs is down. Opinion surveys suggest that Americans are growing more concerned about personal responsibility, as conservatives have narrowly defined that term.

Nevertheless, cheating is up. Cheating is everywhere. By cheating I mean breaking the rules to get ahead academically, professionally, or financially. Some of this cheating involves violating the law; some does not. Either way, most of it is by people who, on the whole, view themselves as upstanding members of society. Again and again, Americans who wouldn't so much as shoplift a pack of chewing gum are committing felonies at tax time, betraying the trust of their patients, misleading investors, ripping off their insurance company, or lying to their clients.

Something strange is going on here. Americans seem to be using two moral compasses. One directs our behavior when it comes to things like sex, family, drugs, and traditional forms of crime. A second provides us ethical guidance in the realm of career, money, and success.

The obvious question is: Where did we pick up that second compass?

History offers some initial clues to this puzzle. Cheating is not a new problem in the United States or anywhere else. It has existed in nearly every human society.

In Ancient Greece, the Olympic games were rife with cheating. Athletes lied about their amateur status, competitions were rigged, judges were bribed. Those caught were forced to pay fines to a special fund used to erect statues of Zeus. Greece ended up with a lot of statues of Zeus. In ancient China, there was frequent cheating to get admission to the civil service. Test takers sewed pockets into their garments for smuggling crib notes and resorted to other creative deceptions. The persistence of cheating on civil service tests was especially impressive given the penalty imposed on those caught: death.

The United States, for all of its moral preoccupations, distinguished itself early on as a natural home to the cheating impulse. Suspicion of authority was part of the fabric and fable of American life from the Republic's earliest days. A search for personal liberty is what brought many to the New World after all, and frontier culture and Jeffersonian suspicions of centralized power nurtured this mind-set. Later, America embraced the rawest form of industrial capitalism in the world. Amid rough-and-tumble business competition and lax regulation, a certain level of lawlessness became part of economic life. An "anything goes" mentality thrived in a country where everyone supposedly had a shot at success—and where judgments of personal worth centered heavily on net worth. As the great sociologist Robert Merton once said, putting his finger on an ugly paradox: "A cardinal American virtue, 'ambition,' promotes a cardinal American vice, 'deviant behavior.'"[3]

During the Gilded Age in the late 1800s, America's new industrialists waged vicious battles as they built, and fought over, the engines of economic growth: railroads, steel mills, oil refineries, coal mines, and banks. These titans of industry cheated each other, they cheated and destroyed their smaller competitors, and they cheated consumers. The tycoon Cornelius Vanderbilt summed up the ethos of the day in a warning delivered to a business adversary who had swindled him: "You have undertaken to cheat me. I will not sue you, for law takes too long. I will ruin you."[4]

The political and cultural milieu of the Gilded Age was permissive of the abuses by the new capitalist overclass. Staggering inequalities of wealth separated America's industrial elite from average Americans. Money dominated a corrupt political system, while laissez-faire notions of individualism were widely embraced. Many respectable civic leaders and intellectuals openly espoused the notion that some classes of people should dominate others.

The 1920s was another notorious period of cheating. The boom of this decade was accompanied by many of the same conditions that defined the Gilded Age. Economic inequality widened dramatically, reaching its highest point in recorded U.S. history in 1929, when the top 1 percent of families controlled nearly half of all household wealth. The federal government turned quiescent, presided over by caretaker Republican presidents who were more interested in golfing than in regulating business. Sinclair Lewis's complacent suburban protagonist, Babbitt, stands as a memorable symbol of the time, but not nearly as memorable as F. Scott Fitzgerald's decadent characters—characters morally suited to an era in which the most powerful people in society were focused on money and cheated to get ahead.

The cheating of the 1920s did not equal the rogue excess of the late 1800s, but in many ways it was more widespread as the ranks of the affluent expanded in America and new venues for cheating emerged. A massive land boom in Florida triggered myriad swindles, as entrepreneurs and investors traded worthless tracts of swampland and jockeyed for a quick profit. The income tax, which Woodrow Wilson had justified partly by the war, was the focus of mounting resentment, and tax evasion became common in the 1920s, especially among the rich. Modern accounting, a relatively young field, was harnessed to facilitate corporate frauds and shelter personal fortunes from tax collectors. Professional baseball was rocked by scandal when it was discovered that the 1919 World Series had been fixed. Prohibition was widely flouted by ordinary citizens and spawned a huge underground economy.

The stock market crash of 1929 brought the curtain down on the go-go '20s, ushering in a more earnest climate. That mood endured amid a string of national emergencies: World War II, the Cold War, the Civil Rights movement, and the turmoil of the '60s. Also, inequality fell during this period. Industry was forced to share power with strong labor unions and activist government regulators. Working-class Americans saw their wages rise steadily from the '40s through the early '70s, and "equity norms" helped to place caps on what CEOs could earn. Cheating scandals involving executives, athletes, lawyers, accountants, students, and so on can be found throughout these years. But from today's vantage point, the moral tone of American life then seems sober and almost innocent.

That tone began to change in the 1970s. The individualism of the '60s turned toxic as it was stripped of its initial liberating purposes and as positive '60s values like social responsibility—which had counterbalanced the new individualism—lost traction in popular culture. Young people became more cynical and materialistic. The nation drifted without a strong sense of national purpose—stuck, it seemed, in an intractable malaise. Meanwhile, the economic upheavals of the decade—inflation, currency instability, oil shocks, rising foreign competition—mobilized the business community to get leaner and meaner, and to begin a far-reaching assault on government regulation and labor unions. By the end of the '70s the stage was set for a new era of extreme capitalism.

In 1981, after he was sworn in as President, Ronald Reagan pronounced: "Government is not the solution, government is the problem." Elsewhere, Reagan articulated another adage that summed up both his philosophy and the dawning ethos of the time: "What I want to see above all is that this remains a country where someone can always get rich."

Ronald Reagan's election stands as a historic turning point that helped crystallize and accelerate emerging trends in American society. Government activism was out. Making money was in. And over the next twenty years, the ideas and values associated with the free market would reign in U.S. society with more influence than at any time since the Gilded Age. "By the end of 2000," wrote one observer, "the market as the dominant cultural force had so infiltrated society that it is increasingly difficult to remember any other reality."[5]

This seismic change has altered the terms of American life. First, thanks to America's laissez-faire revolution, a focus on money and the bottom line has swept into areas that previously were insulated from market pressures. Partly this has been a good thing. Consumers have more choices and get better service these days, and we have an economy that nurtures innovation and entrepreneurialism. Yet there's been a cost. More people in more occupations are chasing money—or being chased by bean counters.

Second, income gaps among Americans have soared over the past quarter century. When profits and performance are the only measure of success, old-fashioned ideas about fairness go out the window. Lean-and-mean business strategies have conspired with trends like globalization and technological change to ensure huge income gains by well-educated professionals—while many less-skilled workers have been running in place or losing ground. Fewer people also control more of the nation's wealth. In fact, the top 1 percent of households have more wealth than the entire bottom 90 percent combined. Economic inequality has led to striking changes in our society.

- In America's new winner-take-all society there is infinitely more to gain, and to lose, when it comes to getting into the right college, getting the right job, becoming a "hot" reporter, showing good earnings on Wall Street, having a high batting average, or otherwise becoming a star achiever.
- Higher inequality has led to more divisions between Americans and weakened the social fabric—undermining the notion that we're all "in it together" and bound by the same rules.
- Inequality is also reshaping our politics as wealthier Americans get more adept at turning money into influence—twisting rules to their benefit and escaping punishment when they break the rules.
- The dramatic upward movement of wealth to top earners has resulted in less wealth for everyone else. Anxiety about money is rife, even among solidly middle-class Americans.

A third consequence of the market's ascendancy is that government's ability to enforce norms of fair play, serving as a "referee" of competition, has been hobbled. Government watchdogs in many areas were disarmed in the '80s and '90s in the name of privatization and deregulation. Extreme laissez-faire thinking has held, foolishly, that the business world can police itself—that the "hidden hand" of market competition will enforce moral behavior and fair outcomes.

Finally the character of Americans has changed. Those values associated with the market hold sway in their most caricatured form: individualism and self-reliance have morphed into selfishness and self-absorption; competitiveness has become social Darwinism;

desire for the good life has turned into materialism; aspiration has become envy. There is a growing gap between the life that many Americans want and the life they can afford—a problem that bedevils even those who would seem to have everything. Other values in our culture have been sidelined: belief in community, social responsibility, compassion for the less able or less fortunate. The decline of civic life, famously described by Robert Putnam, has both fueled these changes and been fueled by them. Everywhere the collective spirit needed for a vibrant civil society is struggling to survive in an era where shared goals are out of fashion.

WHY HAVE THESE transformations led to more cheating? There are four key reasons:

New Pressures. In today's competitive economy, where success and job security can't be taken for granted, it's increasingly tempting to leave your ethics at home every morning. Students are cheating more now that getting a good education is a matter of economic life and death. Lawyers are overbilling as they've been pushed to bring in more money for the firm and as it's gotten harder to make partner. Doctors are accepting bribes from drugmakers, as HMOs have squeezed their incomes. The list goes on. You can even see this problem among cabdrivers in some cities. As cabdrivers have gone from salaried workers with steady incomes to "free agents" who rent their taxis and have to hustle to make a living, they've been feeling new pressures to pick up and drop off as many fares as possible every day. And big surprise: They're speeding and running more red lights.

Bigger Rewards for Winning. As the prizes for the winners have increased, people have become more willing to do whatever it takes to be a winner. A CEO will inflate earnings reports to please Wall Street—and increase the value of his stock options by $50 million. An A student will cheat to get the A+ that she believes, correctly, could make the difference between Harvard and a lifetime of big opportunities—or NYU and fewer opportunities. A steady .295 hitter will take steroids to build the muscles needed to be a slugger—and make $12 million a year instead of a mere $3 million. A journalist will fabricate sources in his quest to write as many hit pieces as possible—so that the day arrives sooner rather than later when he can command six-figure book deals and get lucrative lecture gigs. A partner at a top accounting firm will keep quiet and go along as a client cooks the books—in order to protect a mid-six-figure bonus pegged on bringing in and retaining clients, not angering them.

Twenty-five years ago, many of the huge rewards being dangled in front of professionals didn't exist in a society with less wealth and a stronger sense of fairness. But in the '80s and '90s we came to live in a society where lots of people were striking it rich left and right—and cutting corners made it easier to do so.

Temptation. Temptations to cheat have increased as safeguards against wrongdoing have grown weaker over two decades of deregulation and attacks on government. Many of the recent instances of greed and investor betrayal on Wall Street, for example, could have been prevented by reforms intended to keep accountants honest—or to ensure the independence of stock analysts, or to stop corporate boards from being packed with cronies, or to keep companies from handing out so many stock options. Reformers tried to enact such measures for years, only to be blocked by powerful special interests and antigovernment zealots. At the same time, federal agencies like the SEC, the IRS, and the Justice Department have been starved of the resources needed to stop white-collar crime. Why not inflate earnings reports if the chances of being prosecuted are next to nil? Why not commit a fraud that nets you $70 million—when a year or two in a Club Fed prison camp is the worst possible punishment? Why not hide your income in an illegal offshore bank account when you know that the IRS is too overwhelmed to bother with you because it actually lost enforcement capacity during the '80 and '90s—even as the number of tax returns increased?

Professional watchdog groups have also been asleep on the job. Why worry about being disbarred for bilking your clients when state bar associations lack the resources or wherewithal to fully investigate much of the misconduct by lawyers reported to them? Why worry about being censured by your state's medical society for taking kickbacks to prescribe certain drugs when those groups are more interested in protecting the interests of doctors than of the general public? Why worry about being thrown out of baseball for using steroids when neither of the major leagues has mandatory drug testing?

Growing temptations to cheat have been all the more seductive given the trumpeted morality of the free market. If competition is good—if even greed is good—then maybe questionable cutthroat behavior is also good. In principle, few Americans embrace the idea that "might makes right." In practice, this idea now flourishes across our society, and much of the new cheating is among those with the highest incomes and social status. The Winning Class's clout inevitably has produced hubris and a sense that the rules governing what Leona Helmsley called "the little people" do not apply to them.

This hubris is only partly founded on the kind of delusions made possible by a culture that imputes moral superiority to those who achieve material success. It is also founded on reality. The Winning Class *can* get away with cheating, if not always then certainly often. And when they do get punished, they often find that it's a cinch to later repair their public image. Rehabilitation in the wake of what scholar David Simon memorably labeled "elite deviance" has become easier in recent decades as bottom-line commercialism has steered the media away from critical inquiry toward a new focus on infotainment, much of it celebrating the accomplishments of the rich and famous.[6]

In short, the Winning Class has every reason to imagine that they live in a moral community of their own making governed by different rules. They do.

Trickle-down Corruption. What happens when you're an ordinary middle-class person struggling to make ends meet even as you face relentless pressures to emulate the good life you see every day on TV and in magazines? What happens when you think the system is stacked against people like you and you stop believing that the rules are fair? You just might make up your own moral code. Maybe you'll cheat more often on your taxes, anxious to get a leg up financially and also sure that the tax codes wrongly favor the rich. Maybe you'll misuse your expense account at work to afford a few little luxuries that are out of reach on your salary—and you'll justify this on the grounds that the people running your company are taking home huge paychecks while you're making chump change. Maybe you'll lie to the auto insurance company about a claim or about having a teenage driver in the house, convinced that the insurer has jacked up your rates in order to increase their profits—then again, maybe you have nothing against insurance companies but the payments on that flashy new SUV you just had to have are killing you and you're desperate for any kind of relief.

In theory, there is limitless opportunity in America for anyone willing to work hard, and it seemed during the boom of the '90s that everyone could get rich. The reality is that a lot of families actually lost ground during the past two decades. Middle-class Americans are both insecure and cynical these days—a dangerous combination—and many feel besieged by material expectations that are impossible to attain. It shouldn't come as a surprise that more people are leveling the playing field however they see fit.

THE PAGES THAT FOLLOW offer a journey through many different professions and unfold a host of scandals large and small. They do not cover every angle of what is, unfortunately, a very big problem in the United States. There is much more to be said about cheating than what I say in this book. For example, the role of technological change in increasing cheating could be explored further, while comparisons between the United States and other countries could help illuminate which kinds of conditions in a society are most to blame for cheating. While cheating in the United States appears to be associated with the rising influence of free-market forces, cheating is also pervasive in many countries that extensively regulate the market and where individualism and materialism are less pervasive.

In the end, this book is more a work of social criticism than social science. There is little to no good data that compares most types of cheating today with past points in time, and, in any case, cheaters usually won't talk openly about their actions and motives. What I've done is test my hunch of what is going on by using the information that is available. My hope is that others will drill deeper into some of the issues I've raised to unearth new evidence and insights. I hope that this will be the beginning of a conversation, not the last word.

And a real conversation about cheating is exactly what we need right now. Widespread cheating is undermining some of the most important ideals of American society. The principle of equal opportunity is subverted when those who play by the rules are beaten out by cheaters, as happens every day in academics, sports, business, and other arenas. The belief that hard work is the key to success is mocked when people see, constantly, that success comes faster to those who cut corners. The ideal of equal justice under the law is violated when corporate crooks steal tens of millions of dollars and get slapped on the wrist, while small-time criminals serve long mandatory sentences.

And the victims of cheating aren't just amorphous national ideals but also real people: The elderly pensioner who must get by on less every month because WorldCom cooked its books and leading pension funds lost billions. The musician who doesn't get a second record deal because his first album was infinitely more popular with music swappers than with music shoppers. The patient with a special health problem whose doctor gets a cash payment for suggesting a clinical drug trial she is unfit to take part in. The low-income mother who stays up at night keeping rats away from her children because her landlord falsely certified to city inspectors that he had fixed the holes in her apartment. The average American taxpayer who pays some $3,000 a year more than he should because of other people's tax evasion.

Since the days of Tocqueville, foreign visitors have often marveled at the easy trust among Americans who dealt with each other on an equal footing. While America never has been the fabled classless society of myth, it's managed a close approximation of this myth at different moments. We're not in such a bright moment right now. Instead, we're starting to feel like a corrupt banana republic—one of those places where a rapacious oligarchy sets the moral tone by ripping off the entire country and those below follow suit with corruption of every conceivable kind.

Yes, America needs a serious debate about cheating, but be forewarned: the cheating culture will not be dismantled easily. In many places the root causes of cheating have receded into the background and self-perpetuating dynamics have taken hold, generating their own imperatives for dishonesty. When cheating becomes so pervasive that the perception is that "everybody does it," a new ethical calculus emerges. People place themselves at a disadvantage if they play by official rules rather than the *real* rules. What competitive high school student is willing to tolerate a lower class ranking than other students who are cheating? What law-firm associate hoping to make partner wants to honestly bill the hours she worked if she knows all the other associates are padding their hours and appearing more productive? What pharmaceutical sales director pushing a new prescription drug will forgo showering doctors with expensive gifts when he knows that such bribes are being doled out by competitors pushing rival drugs? What car salesman wants to admit to customers that the next shipment of the hot new model won't be in for eight weeks when all the other salesmen are saying three weeks and making more sales?

Many of us won't give in to pressures to cheat even when we perceive that everybody else does it. We'll study harder to outdo the cheating students, or train more fanatically to beat the athletes who use drugs, or simply make a point of living our lives in more ethical arenas. But all this means playing by our own rules rather than the prevailing rules, which makes life harder in the process. It means being a hero. It's easier to just go along with the cheating culture. And often, when you're deep inside a system where cheating has been normalized, you can't even see that there are choices between being honest and playing by corrupt rules.

In those areas where cheating is not yet widespread, an altogether different calculus prevails. Cheating can be very tempting. It becomes a secret weapon that really can get you ahead. Most people feel uncomfortable gaining an unfair advantage, but many will put aside their qualms if they are under enough financial pressure or if the carrot dangling before them is large enough. People are also more likely to set aside such qualms if society is giving them permission on a larger cultural level.

Yet there is no reason that this erosion of American morality and the tarnishing of core Americans ideals must go on indefinitely. We are, after all, a nation of self-improvers. I suggest a number of steps to reduce cheating in the final chapter. These include creating more broadly shared economic opportunity in U.S. society so that everyone has a chance to get ahead, strengthening democracy so that we all have a more equal say in how the rules are made and so that rules get fairly enforced, and bolstering the vitality of community life to reduce the divisions among Americans that have grown up in recent years.

The final chapter also calls for a sustained assault on entrenched cheating in different institutions and for a new commitment to teaching future generations of Americans to be more ethical. The proposals here include stronger codes of ethics in businesses, universities, and other parts of our society.

No cultural moment in America lasts forever. The one we have been in for the past quarter century—call it the Market Era—may seem permanent, but it is not. History hasn't ended in the United States or anyplace else.

The trick at moments like these is to make history move faster and change arrive sooner.

CHAPTER TWO

Cheating in a Bottom-line Economy

PICTURE YOURSELF AS AN AUTO MECHANIC WHO WORKS FOR a national chain of repair shops. You never went to college, but you got some vocational training and you're making about $30,000 a year. In the old days, when your dad was your age, a mechanic's paycheck would have been big enough to afford the down payment on a modest house. It would have been big enough to allow your wife to stay home with the kids, and maybe even big enough to afford a boat.

Those days are long gone. Your paycheck just gets you by, and nothing more. On the other hand, working for a chain ensures a fair level of security. Customers flock to the chains because they trust them more, and heavy advertising and promotions keep drivers coming back. The work is steady and straightforward. Cars come in; you check them out and tell your manager what needs to be done. Then you fix them. When the day ends, the job doesn't come home with you.

Then abruptly something changes. A new set of marching orders is handed down to the repair centers from the chain's corporate headquarters. You're informed that all mechanics and managers will henceforth have their base pay sharply reduced and will have to make up the difference every month depending on how much work they perform. Once you and the other mechanics do the math, you realize that headquarters is expecting everyone to work harder for the same amount of money. Management also makes it clear that anyone who doesn't meet the quotas may be fired. The scant good news in all of this is that if you work really hard, a special incentive in the new system will allow you to make more money than you did before.

Your job becomes very different. You work harder during the day and worry more at night about whether you'll hit your monthly quota. You also face some dicey choices that didn't exist before. Whenever a customer brings a car in you can proceed in one of two ways: The right thing to do is to take a look and say, honestly, exactly what the car needs. This is your instinct and what you'd prefer to do. After all, you consider yourself an honest person. You've never been in trouble with the law and, except for speeding in your lovingly restored '75 Mustang, you pretty much play by the rules. The problem is that doing the right thing has suddenly gotten a lot harder. You'll suffer a pay cut and risk losing your job if you don't meet your quota—and there's a nice carrot dangling in front of you if you do rack up a lot of extra work.

So gradually you start handling things differently. When a car comes in you take a look, say what the car needs, and then also call for a couple of other repairs that weren't necessary. Maybe the car only needs new brakes, but you say that the struts are also shot and should be replaced. Maybe the sputtering of the engine only requires a good tune-up; but you say that a new carburetor or fuel injection system is in order. Ninety-nine percent of customers don't know the first thing about how cars work, so the chances of getting caught seem next to nil. Also, the other mechanics are doing the same thing, and your manager actually encourages the practice. He's under even more pressure than you are.

Pretty soon, everyone at your shop is cheating customers nearly every day and nobody is giving it much thought. It's just the way things are done now that everyone is hustling so hard to keep their head above water. And your shop is nothing unusual. Some estimates of the cost nationwide of auto-repair fraud run as high as $40 billion a year. Maybe none of the customers at your shop ever catch on and this new way of doing business is never challenged. Or maybe state law enforcement agencies get tipped off to the fraud and they mount a sting operation—as happened with Sears.

During the 1990s, Sears was caught up in a huge scandal regarding its automotive repair chain. The circumstances paralleled the hypothetical situation just described. The scandal underscores how a new and intense focus on the bottom line in American business often spawns more cheating, eroding the integrity of people who would rather be playing by the rules.

Twenty years ago, Sears, Roebuck, & Company reigned as one of America's most well-known and beloved large retailers. For over a century it had prospered in good times and weathered hard times. Sears built more than brand loyalty; it shaped America's popular culture. But times became tougher for Sears in the 1980s as it faced challenges from Wal-Mart and Kmart. Then came the economic downturn at the end of that decade, and Sears confronted a serious falloff in earnings as the 1990s began.

In earlier business eras, like the '50s and '60s, the ups and downs of major companies did not have the consequences they do today. Corporations were firmly controlled by their top executives and often employees had a strong voice through labor unions. A company experiencing hard times could decide internally how to deal with the challenge.

All this changed in the 1970s and 1980s, as large outside investors and Wall Street came to wield increasing power over publicly held companies. A company's stock price became the all-important indicator of success, and businesses were expected to do whatever it took to improve their bottom line—now defined not as healthy performance over a period of years but as strong earnings every quarter of every year.

This meant that a company's leadership no longer had much leeway in handling problems. When earnings fell off, intense pressure mounted to take drastic action. Which is exactly what Sears did as its earnings slid. It announced that it would cut 48,000 jobs and it also instituted a new compensation system for those employees who were spared the ax.

The Sears auto repair chain was the largest in the country at the time, servicing 20 million cars annually. Within a year of the new policies taking effect, complaints from customers were rolling in to consumer groups and state watchdog agencies. People complained of getting billed for repairs they didn't want or need, and of pervasive dishonesty at the Sears repair centers. Sears became the target of official investigations in forty-four states and eighteen class action suits were filed against the company. "We had an all-American relationship with Sears," said Michael J. Stumpf, one angry customer. That relationship ended when Stumpf and his fiancée got stuck with a $650 bill for what should have been an $89 strut job. "Trust shaken is not easily gained back," Stumpf said.

The Sears mechanics and sales staff spoke out about how the new bottom-line focus had forced them to make unpleasant ethical choices at work. One mechanic talked of being "torn between moral integrity, losing my job, and trying to figure out how to work all this out" and others talked of the "pressure, pressure, pressure to get the dollars."

The evidence unearthed by investigators found nearly identical reports of cheating at one Sears auto repair shop after another. It was almost as if Sears's high command had constructed the perfect natural experiment around personal ethics. Want to know what happens when bottom-line practices put ordinary people under intense financial pressure—but provide them with a cheating option that can relieve that pressure? The answer is not very surprising: They will sacrifice their integrity before their economic security.

Sears made a point of getting rid of its commission system after the scandal. It also paid tens of millions of dollars in settlement fees and restitution related to this case. But thanks to a high-priced legal team—including six former state attorney generals—the financial penalties were relatively small and nobody went to prison. Sears CEO Edward Brennan conceded only that "mistakes did occur" and the settlement did not require that Sears admit any wrongdoing. Given this slap on the wrist, it might come as no surprise that cheating continued at Sears Automotive. Lawsuits filed in 1995 and 1999 accused Sears of bilking car owners out of $400 million for tire balancing that was never performed, while another investigation in the late 1990s focused on sales of defective batteries. (Sears ended up agreeing to pay $62.6 million in 2001 to avoid criminal prosecution over the batteries case.) Most recently, in 2002, an undercover sting operation in New Jersey found 350 examples of fraudulent practices at six Sears auto repair centers, mainly the performance of unnecessary repairs.[1]

Auto repair scandals, it must be said, have not been the only problems dogging the company. Sears also faced a major scandal in the 1990s around its credit card business that involved flouting legal protections for holders of Sears cards who had declared bankruptcy and couldn't pay off their debts. Sears paid a $60 million criminal fine in 1999 to settle that case, as well as another $220 million in civil fines and restitution.

The shadow that now hangs over Sears's reputation is rather remarkable, given the company's history. Sears had cultivated a successful love affair with the American consumer for a century. It got through the Depression and many smaller economic downturns without betraying that trust or experiencing serious scandals. And then, during a period of less than fifteen years, Sears squandered public trust, incurred numerous criminal charges and multiple lawsuits, and paid fines and settlements that ultimately totaled over $2 billion. These problems didn't arise because of uniquely bad management or a suicidal impulse on the part of Sears's top executives.

The company and its employees were simply changing with the times.

THE WHITE-SHOE CORPORATE law firms of Manhattan wouldn't seem to bear even a passing resemblance to Sears auto repair centers. These firms operate in a rarefied strata of money and power—managing the legal affairs of some of America's largest companies. They are filled with well-compensated professionals bound by oath to abide by a rigorous code of ethics. And yet life at such firms often entails an ongoing clash between integrity and economic well-being not unlike what occurred in the Sears auto centers. Ordinary people are subjected to intense pressure and high financial stakes—all the while enjoying the option to cheat. And how do many lawyers behave under these circumstances? Not any better than the Sears auto mechanics did.

"Ordinary people" is admittedly not the best description of lawyers at top corporate law firms like Cravath, Swaine & Moore, or White & Case, or Skadden, Arps. Yet the term is not as off the mark as one might think. A few decades ago, the young lawyers who joined such firms every year as new associates were almost exclusively white men from privileged backgrounds. This is not the case today. Most of these firms recruit for talent, with an eye toward diversity, and they have plenty of choices—legions of people graduate from law school every year. Many of the very best of these grads come from middle-class or immigrant families; nearly half of the grads are women and a fifth are minorities.

Despite the knocks that law has taken in recent years, it is still viewed as a reliable path to the Winning Class. The field is also remarkably meritocratic compared to the past. To have a shot at entrance to the best law schools—and thereafter the top law firms and serious money as a lawyer—all a college grad needs are excellent grades, very high LSAT scores, and a willingness to incur huge student debts. Thousands of hardworking young people from a variety of backgrounds meet these requirements, and the entering crop of first-year associates at many top law firms today pulls from a broad cross section of America.

To be a young law associate at a white-shoe New York firm is both a great privilege and a grueling experience. Associates enter a world of dramatic disparities in pay and power, and they are subjected to incredible stress. First-year associates start with a pay package that seems impressive: $125,000 a year and a signing bonus as high as $40,000. This pay becomes less impressive when one considers that associates can spend up to eighty hours a week in the office, the average one-bedroom apartment in New York City costs about $2,200 a month, and the typical law school grad is carrying over $80,000 in student loans. The entry-level pay is not why people join the corporate law world. They join with the hope that someday they will be able to make partner. The average partner at Cravath took home nearly $2 million in bonus pay in 2002. Partners at Wachtell, Lipton, Rosen & Katz took home even bigger bonuses.[2]

But one of the first things associates at top firms learn is that their chances of ascending to partner are next to nil. Only a tiny fraction of associates have any prospect of making partner. The age is long gone when a brilliant young associate could pretty much bank on being elevated to partner. "People will get rejected who are extraordinary lawyers, who have worked long and hard and then get screwed," says a former associate who worked at one of New York's most prestigious firms in the late 1990s. "It's awful and there's awful pressure. For the people really gunning for partner there is a pressure, in addition to working inhuman hours, of trying to beat odds that are incredibly difficult to beat."

Associates at top firms put in insanely long hours in reality—and often longer hours on paper. "You're supposed to bill in tenths of an hour—six minute increments," explains the former associate. "But nobody sits there with a stopwatch. You always round up. If you worked seven minutes, you bill twelve; one minute, you bill six.... Everybody does it at some point."

The anonymity of corporate law makes the corner cutting less troubling. Unlike the old days, there's little loyalty between law firms and clients. "You're part of a sea of lawyers, you have no contact with anyone related to the client," says the former associate. "Why not err on the side of a higher [billing] number? There's no accountability at all for it in that situation. It never comes back to you in any personal way. All these firms charge by the hour. There's an incentive to overbill for work not done or for things that are unnecessary. There's an incentive not to staff things efficiently. It's true everywhere."

One might think that the legal fees of corporate law firms, which can top $500 an hour, would include having legal briefs and other documents typed up and photocopied. One would be wrong. A new term entered the legal lexicon in the 1990s: "Skaddenomics." It referred to the increasingly common practice of billing clients for basic office expenses like faxing and long-distance calls at the kind of outrageous rates one might expect from a five-star hotel. Skadden, Arps happened to be among the worst offenders in this regard, and was singled out by the *American Lawyer* in an investigative article on the practice. But many other firms engaged in the same practice. In the mid-1990s, Cravath was charging clients $1 a page to send faxes, not including the long-distance charges. It charged 15 cents a page for photocopying, up to $91 an hour for librarian time, $50 an hour for word processing, and $40 an hour for proofreading. The *American Lawyer* estimated that Cravath generated $6.6 million in revenue just from its secretaries. None of this gouging was in any way out of the ordinary across the field of corporate law. "People were dreaming up schemes," one law firm administrator admitted to the *American Lawyer* about the practice of Skaddenomics. "If you could measure it, you charged for it." These abuses, which violate the rules of professional conduct outlined by the American Bar Association, typify the bottom-line revolution that has swept through corporate law.[3]

Michael Trotter, a longtime corporate lawyer in Atlanta, summed up that revolution this way: "The years 1960 to 1995 witnessed the transformation of corporate law firms in America from small, dignified, prosperous, conservative, white male professional partnerships dedicated to serving their clients and communities into large, aggressive, wealthy, self-promoting, diverse business organizations where money is often valued more highly than service to clients or community."[4]

Trotter began his career at an Atlanta corporate law firm in 1960. The firms in Atlanta were considerably smaller than New York firms, and the majority of lawyers in these firms were partners—law firms had not yet learned the trick of working to death armies of young associates while reaping all the profits. Lawyers worked eight- or nine-hour days, and on Saturday might put in half a day. A typical workload for Atlanta lawyers was 1,400 or 1,500 billable hours. Trotter's firm was ahead of its time in that it kept track of the billing hours, although it did not keep track of the hours of individual lawyers or expect the lawyers to meet a quota of hours each year.

In the 1970s and 1980s, the firms started tracking the billable hours of every lawyer and holding them accountable for their time. A new bottom-line logic held that the hours each lawyer billed reflected his or her contribution to the firm's profitability. Also, as Atlanta law firms were forced to pay higher salaries to attract talented young associates, they sought to maintain profitability by squeezing more time out of all of their lawyers. "By the early 1990s most major business practice firms in Atlanta expected their lawyers to record 1,800 to 1,900 billable hours a year," Trotter says. Life at New York firms was even more grueling, as billing expectations shot up to over 2,200 hours in most places.

Before the 1980s, for nearly a hundred years an implicit social contract had governed major firms. It was known as the "Cravath model." Truly top-notch associates who worked hard had a good chance of making partner after eight or ten years. Partners were guaranteed tenure for the rest of their careers and, in turn, would be loyal to the firm. Profits were divided evenly among the partners, with senior partners commanding a larger share.

The contract fell apart in the 1980s because of competitive pressures and outright greed. Partners realized they could keep a lot more money for themselves if they strictly limited the number of partners. Increasingly, partners pulled up the ladder to riches and job security behind them and shoveled more work onto enslaved young associates, as well as to cheaper paralegals. The life of the associate turned into a nightmare of killer hours and intense anxiety.

Starting salaries went up dramatically for associates during the 1980s and 1990s—soaring from $18,000 in 1977 to $85,000 in 2000 and then higher—but this money often purchased little in pricey urban areas and a terrible toll was exacted in the billable-hours requirements. The lucky few lawyers who do make partner aren't immune to the stresses of the new bottom line. The old seniority system of equity sharing was rejected in the '80s at most firms in favor of an "eat-what-you-kill" approach. Partners who brought in new clients made big money; those who didn't were marginalized. At top law firms, partners began gunning for colleagues who weren't pulling their weight, and the unquestioned security that came with being partner turned into a thing of the past in many places. "If you're going to exist in this competitive environment," a Boston attorney told a reporter in 1987, "you've got to root out less productive seniors."[5]

These harsh changes in the law took place in the very best of economic times. Money rolled into big law firms as never before in the '80s and '90s. Corporate law partners lived like kings in the '80s, even if they were mere servants to the Masters of the Universe on Wall Street. The last few years of the '90s saw an even greater increase in law firm profits. In 1995, the firm of Wachtell, Lipton, Rosen & Katz posted the industry's highest profits per partner: $1.4 million. Just five years later, Wachtell was again number one, only profits per partner had more than doubled to $3.2 million. Partners at many other top firms also saw their profits double during this period.[6]

The combination of a torrent of money and the legal profession's new bottom line has spawned more cheating. Lisa Lerman is one of the nation's leading experts on corruption in law. She is director of the Law Public Policy Program at Catholic University in Washington, D.C., and her research specialty is candor and deception in the legal profession. She has extensively examined overbilling and other abuses at major law firms, interviewing scores of lawyers. Lerman says that very few lawyers can actually meet the common billing requirements these days of 2,200 to 2,400 hours.

"What I've heard over and over again is people say that they can't bill more than 1,700 or 1,800 hours a year honestly, and their bosses remonstrate them. Everyone knows who's billing the most, and it is not always the one who's working the most. The ones who are willing to play with the numbers are more likely to achieve their goals." Lawyers who don't meet their billing requirements place year-end bonuses at risk and, in hard times, are more likely to be downsized.

So how do so many lawyers manage to bill the hours? They simply make up the numbers, according to Lerman, William Ross, Macklin Fleming, and many other veteran observers of the legal profession. They round their time upward and forget about the many things they did all day that weren't billable, like arguing with their wife on the phone about why they were going to miss dinner yet again. Says Fleming about the liberal allowances more and more lawyers make on their time sheets: "This development may be described as the new math."[7]

New associates at corporate law firms find themselves sprawling down a slippery slope as they learn the true rules that govern their universe. "Let me tell you how you will start acting unethically," wrote a law school dean, Patrick J. Schiltz, in an article addressed to law students. "One day, not too long after you start practicing law, you will sit down at the end of a long, tiring day, and you just won't have much to show for your efforts in terms of billable hours. It will be near the end of the month. You will know that all of the partners will be looking at your monthly time report in a few days, so what you'll do is pad your time sheet just a bit. Maybe you will bill a client for ninety minutes for a task that really took you only sixty minutes to perform. However, you will promise yourself that you will repay the client at the first opportunity by doing thirty minutes for the client for 'free.' In this way, you will be 'borrowing,' not stealing. And then what will happen is that it will become easier and easier to take these little loans against future work. And then, after a while, you will stop paying back these little loans.... You will continue to rationalize your dishonesty to yourself in various ways until one day you stop doing even that. And before long—it won't take you much more than three or four years—you will be stealing from your clients almost every day, and you won't even notice it."[8]

One of the fastest-growing sectors of legal professions these days is legal auditing: lawyers reviewing the bills of other lawyers. The president of a small legal auditing company in St. Louis explains her efforts to stop overbilling as an uphill battle. "It's just the culture within a large law firm," she says. "Before there were a lot of senior people doing it, now new people are doing it as well. It's survival to pay their bills.... When things are going well it goes on. When things are going poorly it goes on." This auditor has been reviewing legal bills for the last twelve years, but she knows things slip by all the time. "Auditors can't see everything."

The varied forms of overbilling complicate the cheating problem. Beyond the simple padding of hours, law firms engage in other abuses: overstaffing, by putting four or five lawyers on a project when two or three would suffice; overloading, by doing unnecessary work for a client; and overqualifying, by assigning partners at a premium rate to jobs that associates can handle. Catching each of these tricks requires an expert eye.

Most legal bills are never audited at all, and some are barely looked at. "You tend to know pretty early on how aggressive the client will be about their bills," says the former associate at Cravath. It's easy for lawyers to get away with outrageous things, and legal experts say that law firms may deliberately pad the bills of clients that are known to be less vigilant. The general laxity around bills is the only way to explain some of the charges that lawyers have tried to slip by their clients. Leona Helmsley once found that her lawyer had billed her for a forty-three-hour day. A lawyer named James Spiotto achieved brief notoriety in the 1990s when he revealed he had billed 6,000 hours a year for four years straight. He claimed his problem was not dishonesty but workaholism. A Los Angeles health-care company found that three different lawyers working on a project had submitted bills for more than fifty hours a day. An Annapolis lawyer named Edward Digges, Jr., a Princeton grad with a 300-acre Maryland estate to maintain, billed a client for $500,000 of nonexistent work and $66,000 for Lexis research that actually cost $394. In one of the biggest overbilling scandals of all, the Federal Deposit Insurance Corporation estimated in the early 1990s that it had been overbilled by as much as $100 million by the private law firms it used for outside contract work. Overbilling was found at nearly every firm the FDIC audited. One lawyer named William Duker overbilled the FDIC and the Resolution Trust Corporation by $1.4 million over a two-year period.

And then there's Webster Hubbell, a confidante of the Clintons who was a partner at the Rose Law Firm and went on to be deputy attorney general. Hubbell bilked clients of hundreds of thousands of dollars during the 1980s and early 1990s. He later explained that the fraud occurred after he had "become overwhelmed" by personal debts and other problems.

In 1998, investigators made public a taped telephone conversation between the imprisoned Hubbell and his wife, Suzanna.

"You didn't actually do that, did you? Mark up time for the client, did you?" Suzanna asked.

"Yes, I did," answered Hubbell. "So does every lawyer in the country."[9]

Changes in corporate law since the 1970s, as well as the experience of the Sears auto mechanics, illustrate the power of economic change to shape personal ethics. The Sears mechanics who didn't cheat took a pay cut and risked their jobs. Corporate lawyers who don't pad their hours may place themselves at a disadvantage in terms of making partner or getting a good bonus. In both cases, personal qualms about cheating easily get buried because "everybody's doing it."

"Your entire frame of reference will change," says Patrick Schiltz of what happens to people who spend even a short amount of time within a cheating culture. "You will still be making dozens of quick, instinctive decisions every day, but those decisions, instead of reflecting the notions of right and wrong by which you conduct your personal life, will instead reflect the set of values by which you will conduct your professional life—a set of values that embodies not what is right and wrong, but what is profitable, and what you can get away with. The system will have succeeded in replacing your values with the system's values, and the system will be profiting as a result."[10]

The bottom-line emphasis of laissez-faire ideology is not new in the United States. American culture—with its classless mythology and frontier ethos—has proved uniquely hospitable to seductive market ideas about the power of individuals to shape their own economic destiny. Horatio Alger, a Harvard grad and the son of a Unitarian pastor, made a fortune writing inspirational books about overcoming hardship. The United States lagged decades behind European nations in developing a social safety net that could buffer Americans from the harsh ups and downs of a capitalist economy.

The market's dominance in American life began to ebb with the rise of Progressivism and Liberalism in the first three decades of the twentieth century. Then came the national emergencies of the Great Depression, World War II, and the Cold War, which helped justify an activist federal government and locked into place a new liberal order.

But a resurgence of market ideology began during the economic turmoil of the 1970s. This decade witnessed the rising influence of conservative economic ideas, a fierce backlash to decades of government activism, and a growing view within business that companies needed to get leaner and meaner to cope with foreign competition.

In the field of economics, a new generation of free-market scholars launched an assault on prevailing liberal orthodoxies about the need for active government intervention in the economy. These attacks were spearheaded by the Chicago School of economics, so named because its leading luminaries—including Friedrich von Hayek and Milton Friedman—taught at the University of Chicago. These economists argued that market solutions would create more individual opportunity than government approaches in nearly every case—from broadly promoting the prosperity of the economy to ensuring affordable housing and health care for all. By the 1980s, the ideas of the Chicago School—and students and disciples—had fanned out across the entire discipline of economics.[11]

Free-market philosophy also found fertile ground in the battles over social policy. Conservatives took aim at the safety net that had appeared so belatedly in American society. They contended that poorer Americans shouldn't be coddled by government. The poor didn't need handouts to improve their lot, conservatives said, they needed the discipline of work instead. While liberal ideas emphasized the harsh vicissitudes of a capitalist system—where not everyone could always find work, or work that afforded them basic life necessities like food, housing, and health care—conservatives emphasized the downside of trying to correct for the market's shortcomings. Charles Murray's bestselling attack on welfare, *Losing Ground*, exemplified this line of logic.[12]

More generally, conservatives equated unfettered economic competition with virtue. They took issue with the antimaterialist values of the '60s. A focus on making money wasn't bad, but exactly the opposite: Wealth was the reward for those who worked hard and competed successfully. Wealth signaled virtue, not vice. Inequities of wealth were actually good because they motivated people to work harder.

As conservatives preached laissez-faire ideology in economic forums, policy circles, and within the culture at large during the 1970s, another free-market revolution was gaining steam in business. The malaise besetting corporate America during the downturn of the decade led to the rise of a new breed of corporate leaders and money managers who called for a take-no-prisoners brand of the bottom line in business. This crusade was motivated partly by the foreign competition that knocked a number of industries on their backs in the '70s, helping fuel deindustrialization across the U.S. as larger manufacturers pulled up stakes and moved overseas to lower costs. But the new emphasis on the bottom line was not actually about survival in many companies; it was about increasing efficiency and, ultimately, profits.

Before these changes, back in the 1950s and 1960s, corporate life was pretty laid back. As Paul Krugman writes: "America's great corporations behaved more like socialist republics than like cutthroat capitalist enterprises, and top executives behaved more like public-spirited bureaucrats than like captains of industry." An implicit social contract at most companies bonded workers and management in a web of loyalty. Workers committed themselves to the firm and, in turn, management treated them well. An ethos of equity ruled in what has been called "managerial capitalism," and salaries at the top of the firm were not grossly inflated in comparison to salaries at the bottom. Powerful labor unions helped enforce the terms of this social contract. Many companies also had strong ties to the community, which they saw as another stakeholder, and played a leading role in civic affairs. "The job of management is to maintain an equitable and working balance among the claims of the various directly interested groups," said Frank Abrams in 1951. Abrams was neither a union agitator nor a liberal intellectual. He was chairman of Standard Oil Company.[13]

By the 1980s the social contract in business was largely defunct. The top ranks of business management began catering mainly to just two constituencies: shareholders and themselves. Large institutional shareholders like pensions and mutual funds got new clout in the 1980s as a vast amount of money poured into the stock market. The managers of these large funds were under constant pressure to show results from their stock picks and they leaned on corporate leaders to boost profits—over a period not of many years, but of every quarter. To show these profits, corporate leaders became even more focused on being lean and mean. As "investor capitalism" replaced "managerial capitalism," mid-twentieth-century notions of loyalty within the firm were tossed aside.

Corporate leaders also grew more intent on feathering their own nests. The demise of labor unions in the '70s and '80s meant that workers had less leverage over how company profits were spread around. Social norms about pay equity within companies withered during this time. The top ranks of management could effectively behave however they wanted and pay themselves whatever they wanted—as long as the stock did well and investors stayed happy. "Downsizing" became common, even in good times, and the salaries of CEOs rose dramatically. In the late 1940s, no executive in America made over half a million dollars a year in official compensation. By 1968, after two decades of explosive growth and record corporate profits, CEO salaries had climbed only modestly. General Motors chairman James M. Roche was the highest-paid executive in America that year, making $795,000. But ten years later, during a period of sluggish growth, David Mahoney, chairman of Norton Simon, was the top earner among corporate chiefs, making $3 million a year. By 1988, Michael Eisner, the CEO of Walt Disney Company, made the most of any executive, taking home $40.1 million. Eisner was number one again ten years later, but this time he took home $575.6 million.[14] The new freedom of top executives to funnel more profits into their own pockets gave them an immense incentive to worship the leanest and meanest version of the bottom line, since every new "efficiency" translated directly into personal gain.

Later I'll come back to the other aspects of the market revolution that have fanned cheating, particularly deregulation. The point here is to show how an obsession with competition and profits steamrolled into American life starting in the 1970s. This obsession partly reflected changes in the economy, but it also reflected a shift in values and norms. "The ideal of a free, self-regulating market is newly triumphant," declared Robert Kuttner in his 1996 book *Everything for Sale*. "Unfettered markets are deemed both the essence of human liberty, and the most expedient route to prosperity."[15]

In this climate, behavior that would have been viewed in earlier periods as brazen greed or chicanery became easier to rationalize. Economic life turned more brutish in many places that had previously been insulated from a money-grubbing, cost-cutting, bean-counting, and one-upsmanship mind-set. Everywhere, free-market forces bulldozed long-standing social norms and professional cultures. The bulldozing wreaked havoc in sports, law, business, accounting, medicine, academia, publishing, and other fields. It was the "creative destruction of capitalism" in classic form. Unfortunately, the ethics of individuals and organizations were among the casualties of a resurgent market. As a booming economy pumped rivers of cash into every kind of business organization—and as the carrots for the Winning Class got bigger—bad behavior became more tempting.

MOST READERS WILL hardly be shocked to hear how a new focus on the bottom line has brought out the worst in corporate lawyers. Only car dealers, CEOs, and stockbrokers are trusted less than lawyers.[16]

But who doesn't trust their family doctor? Well, maybe you can't. While public trust in the ethics of medical doctors grew during the '80s and '90s, it shouldn't have. The corrupting effects of greed and the bottom line are also undermining the integrity of the medical profession—as Tawnya Cummiskey discovered in a very unpleasant way.

Cummiskey lives in California. She had been seeing a doctor named Marc Braunstein for five months when he recommended that she purchase a product called BioLean to boost her energy level. BioLean is made by Wellness International Network (WIN), a global company specializing in the sale of health products that are touted as helping people lose weight, prevent disease, and improve their skin. The company operates through multilevel marketing, a business strategy most famously associated with Amway. Multilevel marketing gets ordinary people to act as sellers for a company's products. These individuals, dubbed "distributors," are lured by the promise not just of making money off their own sales but also of making money off the sales of other distributors that they recruit. That's the theory, anyway. In practice, multilevel marketing can often amount to a pyramid scheme. State law enforcement agencies and consumer protection agencies have file cabinets filled with consumer complaints about multilevel marketing.

Tawnya Cummiskey didn't know any of this when she paid Dr. Braunstein $60 for BioLean. Nor did she realize that her purchase included the expectation that she become a distributor, a fact that she learned only when she received a sales kit in the mail from WIN. All she had wanted was the BioLean, a supplement that Dr. Braunstein assured her would improve her health and which he said was only available through WIN. "I was used to following a doctor's advice," she said later. "Doctors have the training, so you've got to respect their expertise." On reflection, Cummiskey decided that she didn't like the BioLean because it made her heart race and because it wasn't a very good deal, given its high price tag. In fact, the whole business around WIN made her uncomfortable.

The doctor didn't easily take no for an answer. The WIN sales literature counsels distributors to aggressively go after what they want, to create their own dreams, to chart their own destiny, and so on. Dr. Braunstein was not about to let a new distributor one row down in his pyramid escape so easily. He pestered her with dozens of e-mails and phone calls to her home and office, trying to convince her to get with the WIN program. He wanted her to buy more supplements and, more important, to start recruiting her friends and family to join WIN as distributors. When Cummiskey refused to go along, Dr. Braunstein began sending her browbeating e-mails. He pointed constantly to the wealth of other WIN distributors and told her to not let her fears hold her back.[17]

This episode sounds unthinkable—isn't your doctor supposed to worry about your health, not his profits? But doctor-patient interactions of this kind are more common than one might imagine—and much more common than they were ten years ago. Braunstein is only one of many doctors pushing Wellness products to patients: to expand its presence in doctors' offices, the company recruits leading physicians to help convince other doctors to get in on the selling action. And Wellness is only one of over a hundred multilevel marketing companies that sell health products. A growing number of physicians are caught up in these pyramid schemes—lured by the promise of extra income.

Doctors involved in the multilevel-marketing companies, in turn, represent only a small fraction of those physicians who are pitching health products to their patients. Reports peg the sale of health supplements by doctors at nearly $200 million in 2001, a tenfold increase from 1997. An estimated 20,000 doctors are now selling supplements from their offices, more than double the number of five years ago.[18]

Doctors have always played an uneasy dual role of caregiver and entrepreneur, and many have argued that selling dietary supplements or skin-care products or customized cancer drugs is not only a logical extension of their business role but an activity that advances patient care. Yet, in fact, product-peddling doctors often violate the rules of their profession—as well as put their patients at risk—to advance their own financial gain.

Consider where Dr. Braunstein veered off into unethical behavior. He stressed to Cummiskey that she could only get the supplements she wanted to improve her health through WIN, but the ethical guidelines of the American Medical Association say that doctors must "avoid monopolistic practices that hold patients captive" and ensure that recommended products are available through other channels so that "patients have a choice." Dr. Braunstein pushed supplements whose therapeutic properties had not been validated by clinical trial, but the ethical guidelines of the AMA say that doctors must ensure that the claims supporting any products they sell to patients "are scientifically valid and are backed up by peer-reviewed literature and other unbiased scientific sources." Dr. Braunstein pushed supplements to Cummiskey in which he had a clear financial stake, and while the AMA's guidelines do not explicitly prohibit in-office sales of health products, they clearly prohibit such conflicts of interest in other areas, such as the sale of medical equipment. These rules arose after the medical profession agreed that too many conflicts of interest occurred when doctors tried to both objectively advise patients and act as salesmen. For obvious reasons caregiving responsibilities and the profit motive don't mix well together. "Having a financial interest in a product, however indirect, sets up an inherent bias," commented one eminent doctor, Wallace Simpson. "It's just human nature; you want the product or approach to work out, and that affects your judgment."[19]

Many in the AMA have pushed for an outright ban on in-office sales of health products, but the doctors who benefit from such sales have blocked this move during ferocious battles at AMA meetings. The best that the AMA has been able to do is to frown strongly on such sales in its ethical guidelines.[20]

So it goes that more and more physicians continue to create "profit centers" for their medical practices, selling health supplements that are unproven and even dangerous. Dr. Gary L. Huber, who directs the Texas Nutrition Institute, has observed that data supporting the safety and efficacy of many of these products are "almost nonexistent, and many of the products are unreliable." He's also said that "almost half of the supplements studied were potentially toxic and that an incredibly high number of them, when combined with prescription drugs, showed that they had the potential for adverse drug reactions." Another doctor, Kathleen Weaver of Portland, Oregon, puts the problem more bluntly, saying that the medical profession is "in danger of going down the drain, one vitamin bottle at a time."[21]

The recent history of the stimulant ephedra is dramatic testament to these fears. Ephedra has been widely touted as a wonder supplement that helps people shed weight and move through life with new energy. Wellness International Network is one of the leading marketers of ephedra. It encourages doctors to sell ephedra and many doctors promote the stimulants to patients as an effective weight-loss treatment. But the February 2003 death of major league baseball player Steve Bechler illuminated the dark side of ephedra, which can interact with caffeine and other substances in harmful ways. At least 117 deaths have been linked to ephedra since 1993, according to the FDA, as well as thousands of episodes of other serious problems, including strokes and psychotic episodes. Definitive medical and scientific evidence about ephedra is not yet available, but enough questions have been raised about its safety that the FDA has banned ephedra-caffeine combinations in over-the-counter medications, and Health and Human Services Secretary Tommy Thompson is trying to ban doctor sales of ephedra and put warning labels on ephedra products.[22]

Not only do doctors continue selling ephedra to their patients at a profit, but some are also fighting efforts to regulate the supplement—and are helped in this noble cause by Wellness International Network.

Even as thousands of doctors have compromised their integrity by pushing products on patients, they have even more brazenly broken the rules of their profession by getting entangled with the pharmaceutical industry, among the largest and richest industries in America today. The scandal around the drug Neurontin shows the variety of creative ways that drugmakers now funnel illegal payments to doctors. It also shows just how far doctors are willing to stray from a professional obligation to patients in favor of financial gain for themselves.

Neurontin was initially approved by the FDA for exactly one purpose: treating epilepsy. Under federal law, drugmakers are only allowed to market drugs for FDA-approved purposes (although doctors can legally prescribe drugs for many non-approved purposes). This puts pharmaceutical companies in a bottom-line bind, since they all know that prescriptions of drugs for "off-label" uses can dwarf sales for approved uses. For example, while Retin-A was approved for the treatment of acne, sales of the drug exploded when many doctors recommended it for reducing wrinkles. Retin-A's maker, Johnson & Johnson, encouraged this use and was eventually sued by the government.

When Neurontin first came to market, its maker, Parke-Davis—a unit of Warner-Lambert, which in turn is now owned by Pfizer—knew that the drug's medicinal properties made it potentially useful for treating a wide range of ailments, from migraine headaches to diabetic pain to bipolar disorder. But it didn't want to pay the high costs of clinical trials to show that Neurontin worked for these ailments. Neurontin was a gold mine—if Parke-Davis could get doctors to embrace a wide range of off-label uses for the drug.

To do this quietly, without getting nailed by the feds, Parke-Davis followed a marketing strategy centered on doctors and the power of peer-to-peer influence. The strategy sought to co-opt doctors in two ways: first by bribing them individually to prescribe Neurontin for off-label purposes, and second by bribing them to promote the drug's off-label uses to their colleagues and the broader medical community. Both strategies worked brilliantly—for a time.

David Franklin, a former research fellow at the Harvard Medical School, was one of the "medical liaisons" hired by Parke-Davis to help promote Neurontin to doctors for off-label purposes. Franklin would tell doctors that Neurontin was great for treating nearly a dozen different conditions—advice that was not grounded in medical evidence. "I was trained to do things that were blatantly illegal," Franklin said later. "I knew my job was to falsely gain physicians' trust and trade on my graduate degree." Senior marketing director John Ford set the tone for these calls in an internal company directive later. "We want their [the doctors'] whole drug budget...Neurontin for pain, Neurontin for everything. I don't want to see a single patient coming off Neurontin before they've been up to 4,800 milligrams a day. I don't want to hear that safety crap, either."

By "safety crap," Ford meant the legitimate concerns doctors might have about using Neurontin for purposes that had not been vetted by clinical trials. Ford needn't have worried too much about these concerns. Franklin found that doctors were willing to go along with his rap about Neurontin, especially when the company threw in various perks. And when it came to perks, Parke-Davis was extremely generous, providing gifts and making assorted kinds of payments to "tens of thousands" of doctors, in Franklin's estimate.[23]

One marketing strategy involved "consultants' meetings" whereby doctors were given an all-expenses-paid trip to a plush resort, plus a cash honorarium, to provide consulting advice to Parke-Davis. What actually happened at these meetings was that doctors attended short seminars that described the various off-label uses of Neurontin—before kicking off to spend the rest of the day at the beach or, in the case of one conference, at Disney World. Parke-Davis didn't even bother recording the "advice" that the doctors were supposedly there to give. Another method for funneling cash or gifts to doctors involved providing "grants" that were ostensibly for "studies" of Neurontin's various uses. Some doctors received payments of $20,000 or more. Money was also funneled to doctors based on how many patients they signed up for clinical trials involving Neurontin.

Clinical drug trials are now one of the most ethically compromised areas of medicine. Hundreds of thousands of patients are recruited across the United States by their personal physicians to serve as guinea pigs for drugs that they know nothing about and which have not been approved by the FDA. Clinical testing has grown in recent years to a multibillion-dollar industry. It involves thousands of doctors and hundreds of drug companies. According to a major *New York Times* investigation, doctors are offered cash payments of up to $5,000 for each patient they find for clinical tests, and their greed may lead them to recruit patients whose health could be adversely affected by involvement in such trials. Many of these trials are poorly supervised, or supervised by physicians who don't have the right specialty. While these cash-for-patient deals are unethical under AMA guidelines, they have become more pervasive in recent years.[24]

The behavior of both doctors and Parke-Davis in the Neurontin clinical trials was shockingly irresponsible. To begin with, the trials weren't even set up to provide useful clinical research but were actually, according to a lawsuit, "a marketing ploy designed to induce neurologists to become comfortable prescribing Neurontin at a far higher dosage than indicated in the FDA approved labeling." Doctors were paid cash for each patient they recruited for such experimentation, and then more cash based on how much Neurontin they prescribed to the patient. As the lawsuit later charged: "The participating physicians were instructed to titrate their patients to higher than labeled dosages of Neurontin to demonstrate that patients could tolerate high dosages of the drug. Rewarding physicians for prescribing Neurontin was another way to increase Neurontin sales because higher per-patient dosages increased the amount of Neurontin sold."[25]

Over a thousand doctors went along with this dangerous experimentation. Many doctors, some from America's top medical schools, also became hired pitchmen for Parke-Davis, taking money to make public claims about the drug that were not supported by evidence. One prominent doctor took in $300,000 giving Neurontin pitches; six other doctors netted at least $100,000. A "speaker's bureau" run by the Parke-Davis marketing department sent these "sales reps" in disguise out to various medical forums where they did not disclose their payments and pretended to be providing unbiased expert opinions. In a similar vein, Parke-Davis hired a firm called Medical Education Systems to draft a series of twelve scientific articles that touted Neurontin for the treatment of migraines, bipolar disorder, and other problems. Medical Education Systems then found doctors at medical schools to sign their names to the articles in exchange for a cash honorarium.

"We were gambling with people's lives," David Franklin later said about Parke-Davis's aggressive strategies. "We were truly experimenting on patients, which put them at risk." While it was Parke-Davis that conceived this troubling marketing effort and put up the money, its ultimate success hinged on the complicity of doctors—many of whom willingly turned their patients into human guinea pigs and lied to their colleagues in order to make an easy buck.[26]

David Franklin's whistler-blower case against Parke-Davis eventually blew the lid off the scams around Neurontin, and, although Pfizer denies that any wrongdoing occurred at Parke-Davis, settlement negotiations were under way between Pfizer and the federal government as of late 2003. The cost of the scam has been significant. Among other things, Neurontin is very expensive, more expensive than alternatives. As a result of Parke-Davis's off-label strategy, in 2002 doctors wrote about 14 million prescriptions for Neurontin worth $1.3 billion, with 78 percent of these prescriptions being for treatments other than epilepsy.[27] These costs are ultimately passed along to consumers through the premiums charged by private insurance companies and also to taxpayers through reimbursements via government health programs. State governments have been particularly angered by the Neurontin off-label scam because it gave doctors incentives to prescribe millions of dollars of the expensive drug to Medicaid patients, ignoring less-expensive alternatives, and states were forced to pick up much of this tab.

While conflicts of interest among doctors are not a new problem—such conflicts have bedeviled medicine for at least a century—it seems that more doctors nowadays are breaking more rules and putting Americans at risk in the process.[28] No one has ever polled doctors about why they turn into sales reps for supplement companies or pushers for drugmakers, or why they conduct dangerous experiments on patients for cash rewards. But if you listen to doctors talk about their professional lives, you'll see that there are plenty of reasons for this, related to both the carrot and the stick.

The stick comes from the new bottom line in medicine. In the 1960s and 1970s many doctors operated as free-spirited entrepreneurs, making all key medical decisions on their own and also making a great living. Today, an increasing percentage of doctors find themselves corralled in managed-care groups where they are essentially employees of huge corporations. Managed care has famously reshaped the experience of being a doctor. To control costs, health maintenance organizations (HMOs) and other managed-care companies have resorted to the same tactics that so many businesses have. They have squeezed their labor force to increase its productivity while trying to keep profit sharing to a minimum. Doctors with managed-care organizations are expected to see a certain number of patients per day and spend the minimum amount of time with each one—sometimes as little as twelve minutes. Doctors also are tightly circumscribed in the kinds of medical treatments they can recommend and many even operate under gag rules that prevent them from telling patients that recommended treatments have been rejected by their managed-care provider.[29]

All of these market-driven changes assault the old-fashioned practice of taking the time to get to know patients and to connect with them compassionately. The antiseptic business language of managed care underscores the new mentality. As one physician wrote in disgust: "Today patients are called 'consumers,' physicians are 'providers,' and health care is a 'product'—all terms of commerce, not of a profession, and not of a humanitarian profession."[30]

These changes have made doctors in HMOs nearly as miserable as associates at corporate law firms, if that can be possible. The managed-care revolution and other changes have also dramatically affected the incomes of doctors. Consider the financial situation of the average young physician. After four years of college and eight years of medical training, 85 percent of new doctors begin their career with large debt loads. During the 1980s the average debt burden for medical school graduates was typically less than $40,000. Today debt loads are approaching the size of mortgages: the average debt of med school grads jumped to more than $60,000 by the mid-1990s, and then soared further, to over $90,000 by 2000. This trend of rising debt has been paralleled by stagnating salaries among physicians as a result of the squeeze from managed care and lagging Medicare and Medicaid reimbursements. While the income of doctors rose rapidly in the '60s and '70s, that increase slowed in the '80s, and then stagnated in the '90s, and incomes even fell for some doctors.[31]

Like other professions, medicine is filled with stark winner-take-all inequities. While a star cardiologist or neurosurgeon can make millions, other doctors struggle to get by. Many heavily indebted young physicians find themselves forever trying to catch up financially. Whether they like it or not, a focus on money becomes central to their lives. "It is little wonder that new physicians pay such close attention to the business and financial aspects of medicine," wrote one doctor. "Debt payments stretch out ahead of them, and management is a daily concern. It was far different for graduates of years ago, who believed that if you worked hard and took good care of your patients and practice, you would do well financially." In a major survey of doctors taken in 2002, over 40 percent said they were dissatisfied with their financial situation, and over half said that they do not expect things to get better in the future. A third said that they would not recommend medicine to young people because of the low financial rewards in the profession. Unhappiness about money comes on top of a much deeper anger among doctors about their loss of autonomy and the degradation of their profession by managed care. In medicine these days, as in other parts of the economy, many people feel that the social contract they signed has been broken—and this perception bodes ill for doctors' commitment to following the rules.[32]

America's health-care system is badly in need of cost-cutting controls, and HMOs pat themselves on the back for imposing such discipline. These pats might be deserved if managed-care organizations were dedicated in principle to ensuring the best medical care. They are not. Their goal is to maximize shareholder returns. HMO profits have been rising steadily over the past few years, and HMOs have responded to this trend by paying their CEOs healthy bonuses. Executive pay in the managed-care field is now two-thirds higher than in other industries. HMO stocks have also been performing better than the market as a whole. "HMO stocks are going up in a down market," commented a September

2002 editorial in an AMA publication, "and doctors appear to be paying the price."[33]

Garnering sympathy for doctors is an uphill battle in an America where the average worker makes $40,000 a year and the average primary-care physician rakes in nearly four times that much. Yet keep in mind that when it comes to money, everything is relative. A urologist doesn't compare himself to the auto mechanic who fixes his Volvo. He compares himself to his father, a doctor who prospered during the golden age of medicine. He compares himself to the doctors whom he went to medical school with, some of whom became star specialists and live in mansions. And he compares himself more broadly to the class of high achievers to which he belongs: to the many professionals his age who make more money than he does in such fields as investment banking, software development, biotechnology, management consulting, and so on. He feels that he paid his dues for a comfortable lifestyle in the form of an eight-year medical education. Yet as a creature of the HMO, he may well find a huge gap between the life he expects and the life he can afford.

Thus he may pay attention to which drugmakers offer him ways to boost his income. He may accept generous gifts, and go with his family to the plush resort on Anguilla or on the whitewater rafting trip down the Grand Canyon. He may listen closely when he hears about other doctors who are making big bucks selling health supplements. Indeed, companies like Wellness International Network explicitly seek to tap into doctors' financial stresses as they try to turn them into multilevel marketers. Doctors, in turn, cite such stresses to justify why they turn their examining rooms into sales centers. "Physicians are trying to survive today," explains Tim Berry, a doctor who was struggling to keep his practice afloat before he became a Wellness distributor. Berry observes that the money he makes selling supplements is money that the "insurance companies can't take away from me." Another doctor who sells Amway products also blames HMOs: "Do you think I went through college and professional school and professional training and all this to have managed care come and tell me that I was worth eight cents on the dollar for a patient each month?"[34]

The carrots that are tempting doctors are extraordinary. A top Wellness distributor can make several hundred thousand dollars a year. Meanwhile, the pharmaceutical industry has very deep pockets. Americans spent $200 billion on medications in 2002, an amount that has soared in the past decade. Capturing the allegiance of doctors and patients for particular drugs involves huge stakes for the pharmaceutical industry, and they spend more lavishly on marketing than ever before. Drug advertising rose an average of 45 percent each year between 1994 and 2000; the pharmaceutical industry spent $2.7 billion on consumer advertising in 2001. In recent years drugmakers have become more sophisticated about going straight to consumers; Americans increasingly ask their doctors about drugs they have seen advertised on television or elsewhere. Ultimately, though, doctors remain the principal gatekeepers to consumer prescription drug choices. And in comparison to the cost of ad campaigns on television or in national magazines, bribes for doctors amount to small change.[35]

The marriage between a pharmaceutical industry with few scruples and doctors under growing financial pressures is among the most unholy of professional matches in America today. Patients are the unwitting losers. Even as we place more trust in doctors than nearly anyone else in our lives, that trust is often betrayed. And, in an ironic twist, many doctors who are doing the right thing for their patients also end up cheating within America's dysfunctional health-care system. When faced with HMOs intent on denying needed treatments to patients, many doctors have few qualms about lying to managed-care providers about the true condition of patients. "In order to get permission to do treatment for serious conditions, you make it sound worse than it is," explained New York psychiatrist Edward Stephens.[36] A survey published in 2000 in the *Journal of the American Medical Association* found that 40 percent of physicians admitted to deceiving insurance companies within the previous year to help patients. In another smaller survey, some 70 percent of doctors said they had exaggerated patients' symptoms so that HMOs would allow them longer hospital stays. Sixty percent said they had altered diagnoses to ensure that treatment was covered. Such pervasive deception results from the judgment of doctors that "gaming the system" is the only way to do their jobs properly, but it also stems from rising pressures on doctors from patients who demand tests or treatments that would not be approved by managed-care organizations.[37]

Many doctors also deceive the big government entitlement programs, Medicare and Medicaid, in order to do right by their patients. The limits on care within these programs have become more harsh and arbitrary as budgets have gotten tighter. This is especially true in regard to Medicaid, where state governments now have wide discretion about how to allocate care to the low-income people who need it. Many states have abused this discretion as part of a zealous get-tough approach to welfare, neglecting those who should be eligible for care. And even when Medicaid is run properly, it does not cover many poor Americans who desperately need care. Doctors in low-income neighborhoods and emergency rooms repeatedly end up in ethical dilemmas: whether to tell the truth on paperwork or to fulfill their professional responsibility.

With budgets tightening further and government health-care

systems under growing strains, these dilemmas will become more common in the years ahead.

EVERYONE AGREES that the landscape of economic life in America has been transformed over the past two decades. Boom times created more consumer demands and myriad new business opportunities. Economic life has also been transformed by a relentless new focus on the bottom line, often to the exclusion of other concerns that once preoccupied business leaders—like having loyal workers, acknowledging seniority, and giving back to the community.

The conventional wisdom is that these changes have been a good thing. We are a far richer nation than a quarter of a century ago, and who can argue with more wealth? Who can argue with competitive markets that bring down prices for consumers while expanding choices? More broadly, who can dispute the dynamism of America's economy over the past two decades? Lean and mean has worked, right? By doing away with old forms of business organizations, we've made the country richer and provided a bonanza of deals for everyone.

Well, yes and no. The United States experienced faster economic growth and bigger income gains in the 1950s and the 1960s than in the lean-and-mean '80s and '90s.[38] Business organizations have become leaner, but mainly for the people at the bottom and in the middle. And, amid the rush to a purer free market—with all its supposed virtue—we never reckoned with how "mean" would affect the quality of life and ethics of professionals or with the unseemly pressures that would be exerted by "lean."

Somehow, we failed to see an obvious downside of lean and mean—that when you put people under pressure and give them a choice of preserving either their integrity or their financial security, many will go for the money.

CHAPTER THREE

Whatever It Takes

In the late 1800s, an English scholar named James Bryce traveled extensively in the United States and wrote a multi-volume study filled with his observations entitled *The American Commonwealth.* Like Alexis de Tocqueville before him, Bryce was struck by how Americans were so much more trusting of each other and so much more generous than Europeans. Bryce attributed this quality, as Tocqueville had, to an ethos of class equality that pervaded American social relations. "People meet on a simple and natural footing," he wrote, "with more frankness and ease than is possible in countries where every one is either looking up or looking down.... It gives a sense of solidarity to the whole nation, cutting away the ground for the jealousies and grudges which distract people."[1]

Does this sound like ancient history or what? If Bryce returned today, he'd find a very different country—a country filled with VIP lounges, personal assistants, chartered jets, exclusive restaurants, luxury hotels, and private golf clubs; a country where superrich celebrities and sports stars reign as demigods, where the wealthy engineer superior looks and health through expensive medical intervention;

a country soaked in poisonous envy spurred on by a $250 billion advertising industry; and a country where millions of affluent people live behind guarded gates.

These unhappy changes—the new class divisions in an America far richer than ever before—are an important part of my story about why cheating has increased over the past few decades.

It's not hard to see how much richer America has become since the 1970s. Just stop in at a Banana Republic clothing store. Over 400 of them are sprinkled across the United States, each one carrying identical upscale clothes. For $78, a man can purchase a pair of classic straight-leg jeans. A hundred dollars buys a fancy pair of wool dress pants. A short-sleeve polo shirt is $42. The ubiquity of Banana Republic stores—or rather, the ubiquity of customers willing to blow $42 on a polo shirt—is testament to the remarkable new affluence of American society. Twenty years ago, Banana Republic consisted of two stores in San Francisco specializing in outdoor clothing, along with a catalog business. After it was bought by the Gap in 1983, it jettisoned any serious connection to outdoor activities and refocused on clothing affluent urban professionals. Other newly expanding chains, like J. Crew, catered to the same demographic. By the mid-1990s, Banana Republic was in every major city in America and was opening over twenty-five new stores a year, expanding in step with America's upper middle class: during the 1990s, the number of households filing tax returns reporting incomes over $100,000 a year quadrupled.

The creation of this mass upper middle class is a historic accomplishment on par with the creation of the mass middle class after World War II. Yet this gain has come at a significant cost. The vast gulf between the top tiers of American earners and everyone else is the most obvious of these costs. How big is this gulf? *Very big.*

To get a sense of these gaps, stick around the Banana Republic store. The people working the cash register and sales floor are typical of the losers in the new economy. Most of them are young entry-level workers without college degrees. As unskilled entry-level workers, these salespeople are doing terribly, in historical terms, and may well be making less money than their parents. If they're making the minimum wage, they're earning a wage that has decreased in constant dollars by over 20 percent since 1979. If they're making more than the minimum wage, they're still not doing well. Between 1979 and 2001, during one of the greatest economic booms of all time, entry-level wages for workers with only a high school degree *fell* by 14 percent. At the same time, employer-provided pensions and health care became less common for these workers. Nearly 30 million American workers—almost a quarter of all working people—earn under $19,000 a year.[2]

The manager at the Banana Republic is likely to be the one lucky employee who actually makes an okay salary and gets good benefits. In fact, Gap, Inc., has some of the best benefits in the retail business, including free therapy for stressed-out workers. But lucky as she is, the store manager may well be doing worse than her blue-collar parents who never went to college. And she's making chump change compared to the professionals who are running the Banana Republic division at the Gap, Inc., headquarters in San Francisco. For example, the senior vice president for product design and development or the senior vice president for marketing are probably making five times the salary of the store manager. These people have firmly made it into the mass upper middle class.

And what about the people at the very top of the Gap's $14 billion empire? Donald Fisher, who opened the first Gap store in San Francisco in 1969, along with his wife, Doris, is worth $3 billion. Millard Drexler, until recently the CEO of Gap, Inc., saw his compensation skyrocket in the late 1990s and walked away with a fortune of over $500 million. Even in 2000, when Gap, Inc., did poorly and saw its earnings fall, Drexler managed to finagle a raise of $25 million. He did this by cutting his bonus while taking many more stock options. The sleight-of-hand earned Drexler a place in the "Pay Hall of Shame" put out each year by Graef Crystal of *Bloomberg News*. Crystal wrote: "Drexler's case illustrates how CEOs often are able to extract a raise in hard times, simply by re-engineering their pay packages in a manner that makes the average shareholder think they took a cut in pay."[3]

The pay structure at Banana Republic mirrors that of the nation as a whole. From the late '70s through the late '90s, the top 1 percent of American households saw their incomes increase by 157 percent. On average, these superwealthy households scored after-tax income gains of nearly half a million dollars. More broadly, all Americans in the top 20 percent of households did extremely well during the 1980s and 1990s. The average after-tax income gains for the top fifth of households increased by over 50 percent.

Families at the bottom of the economic ladder and those on the middle experienced the boom years in an altogether different fashion. For these households, there was no boom at all. In typical fashion, those at the very bottom missed out the most. On average, the incomes of the bottom fifth of households actually *declined* slightly during the supposed boom—going from an average of $10,900 in 1979 to $10,800 in 1997. In the last few years of the 1990s, the bottom tier of earners did see some gains, but these were not great enough to make up for the stagnant wages of the previous twenty-five years.

And what about the people smack in the middle, the fabled pillars of American society? The boom passed them by, too. After-tax income gains by the middle fifth of households through the '80s and '90s were about 10 percent, or $3,400 in new income—less than a tenth of the increases of the top 20 percent.[4]

Wealth gaps have grown even faster than income gaps in America's Banana Republic economy. Forget all the talk of how everybody now owns stock and how we're in some golden age where the "middle class has joined the money class." In fact, we've been hurtling backward in time. The top 1 percent of Americans now holds nearly 40 percent of all household wealth—such as retirement savings, household equity, stocks, etc.—up from 20 percent in 1979. This tiny sliver of the population has more wealth than the bottom 90 percent of households put together. Many Americans lower on the economic ladder have seen their net worth actually decline; the net worth of the bottom 40 percent of households fell by a shocking 76 percent between 1983 and 1998. As for the supposed populism of Wall Street, most Americans missed out on the hyped bull market of the '90s since the lion's share of stocks was—and is—owned by wealthier households. For example, between 1989 and 1997, 86 percent of stock market gains went to just the top 10 percent of households.[5]

Social scientists have long been arguing about the causes of skyrocketing inequality. Leading suspects typically include technological change and globalization. But scholars also blame the new inequality on bottom-line business strategies that have made companies leaner and meaner and have abolished equity norms along the way. Management pays itself bigger salaries while more and more jobs are outsourced to temporary workers without benefits or to contractors who pay poor wages. A rollback of government intervention in the economy has encouraged this trend by keeping the minimum wage low, making it harder for workers to form unions, and reducing government oversight of existing labor and workplace safety laws. Free-market ideologues have also successfully pushed tax polices that favor the wealthy and further concentrate wealth at the top of our society.[6]

Inequality has grown across the industrialized world since the

1970s, but it's more acute here in large part because America has done next to nothing to tackle the problem: There have been no huge new investments in education or job training to ensure that everyone can compete in the postindustrial economy; no consistent government efforts to prime the economic pump to keep labor markets tight and raise wages for those at the bottom; and no major assistance to lower-income families to help them build wealth in the form of homes and retirement savings.

Many say that inequality doesn't matter. It's said that as long as there is opportunity, as long as people can move upward and transform their lives through hard work, inequality is not a social problem we should worry about. "If you drive a Mercedes and I have to walk, that's a radical difference in lifestyle. But is it a big deal if you drive a Mercedes and I drive a Hyundai?" asks Dinesh D'Souza. "Why should people feel aggrieved that the rich are pulling further ahead if they are also moving forward?"[7] If this sounds fine in theory, it's not how things are playing out in reality. For one thing, America's many workers who are without the education or skills to prosper in the new economy have little opportunity to improve their station in life. And mobility is not nearly as great in American society as is commonly imagined. In fact, research shows that the United States now has less economic mobility than some Western European countries such as Sweden.[8]

At the same time, there is mounting evidence that inequality has a variety of poisonous effects on the fabric of American life. As our society has become more divided along income lines, social and cultural gaps between people have increased. The very affluent have retreated from the public sphere, increasingly sending their children to private schools or living in gated communities—while the middle class deals with cuts in services and bad schools. As the rich have used their wealth to grab more influence over public policy, the middle class and the poor feel even more alienated from politics. Meanwhile, even though the economic pie has been expanding, insecurity has actually grown for many workers as their slice of that pie has stayed the same or shrunk—while living costs keep rising.[9]

The yawning gap between winners and losers is also having a lethal effect on personal integrity. In a society where winners win bigger than ever before and losers are punished more harshly—whether by losing a job with benefits or not earning enough money to make ends meet—more and more people will do *anything* to be a winner. This is an absolutely critical point to grasp in understanding the cheating culture, and it is obvious when you think about it. Cheating is more tempting if the penalties for failure are higher, if you're feeling pinched or under the gun, like the Sears mechanics or many lawyers today. It's also more tempting if the rewards for success are greater—if cheating can make the difference between being a multimillionaire or just getting by. When people perceive this kind of choice, they will often kiss their integrity good-bye.

REMEMBER Danny Almonte?

Danny was the Little League champ from the Bronx who awed baseball fans by pitching a perfect game during the run-up to the Little League World Series in 2001. "He's head and shoulders above anybody I've seen come through the Little League system," said one coach. Even though Danny was only twelve—or so he said—there was much talk of Danny's bright future in the major leagues. Danny was such an amazing player that an ABC Sports executive producer described him as "the single biggest story in the history of the Little League."

Danny became an even bigger story when it turned out that he had lied about his age to make himself eligible to play in Little League. Danny's father, Felipe Almonte, an immigrant from the

Dominican Republic, had conspired with Bronx All-Star coach Rolando Paulino to falsify his son's date of birth. The senior Almonte altered Danny's Dominican passport, changing the date of birth, April 7, 1987, to read April 7, 1989. Fourteen-year-old Danny Almonte had violated the most important rule of Little League—that it's for little people.[10]

The revelations triggered an uproar. Even President George W. Bush offered an opinion in the aftermath of the disqualification. "I was disappointed that adults would fudge the boy's age," he said. "I wasn't disappointed in his fastball and his slider."[11] In besmirching Little League, one of the most hallowed icons of Americana, the scandal left many people angry. It was another of those moments in national life, so common lately, when one wondered: Why would anyone do such a thing? How low can people go?

Yet the episode should not have been surprising. More than ever before, the glittering world of sports looms above the lives of poor and middle-class Americans as an escape hatch from the Anxious Class and a ticket to the Winning Class.

Felipe Almonte apparently wagered a great deal on his son's prospects as a future Alex Rodriguez, the extravagantly paid infielder for the New York Yankees who was born to Dominican immigrant parents. Not only did Danny never attend school during his eighteen months as a pitcher for the Bronx All-Stars, but he apparently did not attend much school at home in the Dominican Republic, in part because of his baseball regimen. Felipe Almonte was not unlike any number of other parents these days who raise their children to be stars on the lucrative playing fields and courts of professional tennis, basketball, baseball, and football. And as the rewards for top athletes have risen sharply, more parents have driven more children harder.

Almonte had better motives than most parents. The Dominican Republic is among the poorest countries in the Western Hemisphere, and Dominicans have one of the highest poverty rates of any immigrant group in New York City. Hundreds of thousands of Dominicans live in crowded neighborhoods with poor schools, pervasive unemployment, and high crime—neighborhoods just a mile or so away from some of the wealthiest areas in the United States. Many end up trapped in low-wage, dead-end jobs. Low-skilled immigrants did okay in an earlier America, where factory jobs were plentiful and even a high school dropout had a decent shot at making enough money to be the sole family breadwinner and buy a home. Now it's common for both parents in an immigrant family to work full-time at low-wage jobs and yet barely escape poverty.

In the face of these odds, the world of Little League offered tangible possibilities for Danny Almonte and his family. Little League is no longer a small-town, pint-sized pastime. It can offer a shot at fame and wealth. In 2000, the organizers of Little League baseball began to market the broadcast rights for the Little League World Series for the first time and to make lucrative deals with corporate sponsors. Little League lined up $3 million in sponsors for the 2000 World Series, including major contributions from Honda, Wilson Sporting Goods, and myteam.com. In 2001, around the time that a $7 million stadium project was being completed in Williamsport, Pennsylvania, the number of teams in the Little League World Series doubled as sixteen teams advanced to the finals.

The Bronx All-Stars were a big fish in the expanding Little League pond. David Komansky, then CEO of Merrill Lynch, was a special fan of coach Rolando Paulino and his team. Merrill provided a corporate sponsorship for the team, while executives, including Komansky, attended the games. Boosted in part by the sensational play of Danny Almonte, television ratings for the Little League World Series hit record highs, especially in the New York area.

Little League stardom offered a clear upward path for Danny. A player who excels is guaranteed attention in high school from big league scouts. Big money can lie just around the corner. In 2001, a high school catcher named Joe Mauer was paid a $5.15 million signing bonus when he was recruited by the Minnesota Twins.

The scandal about Danny's age put a damper on his family's dreams of this kind of money, but only temporarily. Danny is living in the Bronx, concentrating on getting through school, and playing baseball for James Monroe High School. The coach there sees big league potential in Danny's pitching. Danny may yet make it to the majors one day.[12]

If he does, he will leave behind one of the most impoverished neighborhoods in one of the most unequal cities in America—and join a profession where men feel poor on $300,000 a year and the inequalities between sluggers and benchwarmers are comparable to the contrast between Harlem and the Upper East Side. He will meet a lot of guys from humble backgrounds similar to his own who are under intense pressure to perform at a very high level and keep their toehold in the Winning Class. Danny will also enter a world so rife with cheating that his own past sins will seem laughable in comparison.

Consider the San Francisco Giants as one example of the winner-take-all market in sports. On opening day in spring 2002, the Giants paid its starting roster of twenty-six players a total team salary of $78.3 million. Almost a fifth of this pie went to one player: left fielder Barry Bonds, who took home $15 million during the season. Over half of the total team salary in 2002 went to five of the Giants' top players. At the bottom of the salary pyramid were seven team members with salaries of $300,000 or less.

The Giants' salary structure is a relatively new phenomenon. In just four years, between 1996 and 2000, the average salary on the team doubled, and the gaps between the highest- and lowest-paid players widened dramatically. This trend reflects the bigger picture in baseball over the past two decades, where salaries along with income gaps have grown exponentially. The average salary of a professional baseball player was $60,000 in 1975, $135,000 in 1980, and $413,000 by 1988. Today it's about $1.5 million—an increase of over 2,000 percent since 1975. A key factor pushing up these averages has been the money paid to top players, which keeps getting higher and higher.

When Barry Bonds first started making over $10 million in 1997, he was the highest-paid player in baseball. A year later, Kevin Brown became the highest-paid player when he signed a contract with the Dodgers worth $15 million a year. Then, in 2000, Alex Rodriguez became the highest-paid player when the Texas Rangers agreed to pay him $25 million a year. During these same three years, the minimum union salary paid to men sitting on the same bench as these megastars only increased from $150,000 to $200,000.

What happens in a sport where top players rake in 50 or even 100 times more than their teammates? Bad things.

Take the career of Bonds. He's a slugger and a star, and he should be basking in glory and serving as a role model to younger players. Instead, he is distrusted and surrounded by controversy. Part of Bonds's problem stems from a cantankerous and arrogant personality. But much uneasiness around Bonds is fed by widespread rumors that his huge salary gains in recent years are the ill-gotten fruits of drug use.

Bonds is widely accused of using steroids starting in the late 1990s to pack on thirty-eight pounds of muscle in just a few years and transform himself into baseball's most powerful slugger. Anabolic steroids are a form of synthetic testosterone and produce hormone levels that help generate more muscle. Bonds's accusers say that his increased power would have been impossible without serious pharmaceutical help, especially given his age. (Bonds was thirty-eight in 2002.) Bonds and his supporters counter that the new muscle came from training and a diet that included Creatine and protein pills. "Barry Bonds could be on steroids," said a leading expert on drugs and sports, Charles Yesalis, "but his power comes from the fact that he has the closest thing to a perfect swing that I've ever seen."[13]

Whatever the truth, Bonds's case vividly illustrates the rewards that await athletes who can bulk up fast. Bonds's bigger muscles and better hitting enabled him to break Mark McGwire's home-run record, slamming seventy-three balls over the fence in 2001. McGwire's own power hitting was fueled by Androstenedione, a steroid banned in the NFL but permitted by the major leagues. Bonds was also able to sign a new contract that nearly doubled his salary over what he was making when he still weighed in at 190 pounds in 1997. Bonds now makes $18 million a year; in the 2003 season, he'll be paid over $35,000 for each time he's at bat.

Steroid use in sports has been around for decades, especially in football and among Olympic athletes. But it is only in recent years that it has become common in baseball. More players are also taking other drugs like human growth hormone (hGH) and amphetamines. *Sports Illustrated* conducted an investigation of the problem in 2002 and reported that "Steroid use, which a decade ago was considered a taboo violated by a few renegade sluggers, is now so rampant in baseball that even pitchers and wispy outfielders are juicing up...the game has become a pharmacological trade show."[14]

The year Mark McGwire beat Roger Maris's record, he stood six feet, five inches tall and weighed 245 pounds. He had twenty-inch biceps and seventeen-inch forearms. Other big-time sluggers are comparably built. And yet, as anyone who has trained with weights knows well, it is not easy to either build or sustain a large mass of muscle over time. To be successful—and to have a shot at astronomical money—today's professional baseball player must do more than push his body to its maximum brawn. He has got to stay at this peak level month after month, year after year. Steroids can help players do this, but they have numerous side effects, including impotence, liver damage, and heart disease.

There have always been enormous incentives for players to take shortcuts, either to get to where they want to be or to stay there. But these incentives increased in the 1990s, as the salaries for reliable sluggers soared into the stratosphere and merely average players made only moderate gains. "A big, big year means a big, big contract," observed Kevin Towers, general manager of the San Diego Padres.[15] A bad year may mean a return to the minors.

While there are no reliable statistics on the pervasiveness of drug use in baseball, a variety of players have offered their own estimates. Yankee star David Wells recently estimated in his memoir that 40 percent of major leaguers use steroids. Jose Canseco, who has played with the Oakland A's, has commented that the use of steroids has "revolutionized" baseball and estimated that 85 percent of players are on the drug. Arizona Diamondbacks pitcher Curt Schilling guesses that the number is between 40 and 60 percent. Said Schilling: "I'll pat guys on the ass, and they'll look at me and go, 'Don't hit me there, man. It hurts.' That's because that's where they shoot the steroid needles."[16]

The few baseball players who have openly explained their steroid use emphasize the financial reasons for risking their health. Ken Caminiti was a third baseman for the San Diego Padres when he began taking banned steroids in 1996 to help overcome an injury. Caminiti did more than just heal. He played as never before—with 40 home runs, 130 RBIs, and a .326 batting average. The year's record earned him the Most Valuable Player of the National League. "At first I felt like a cheater," Caminiti told *Sports Illustrated*. "But I looked around, and everybody was doing it."[17]

Caminiti shares the view that at least half of major league players are on steroids, and he depicts an environment where the use of drugs has now passed a tipping point and become normalized. "If a young player were to ask me what to do, I'm not going to tell him it's bad. Look at all the money in the game: You have a chance to set your family up, to get your daughter into a better school. . . . So I can't say, 'Don't do it,' when the guy next to you is as big as a house and he's going to take your job and make the money."[18]

A minor league baseball player echoed this point when asked why he used steroids. "I've got an easy answer for that. I'd say, You've set up a reward system where you're paying people $1 million to put the ball into the seats. Well, I need help doing that." Players in the minor leagues report pervasive steroid use. Everyone who is there wants out and up. They want to make the majors—and their personal fortunes.[19]

Meanwhile, even the best players at the top of baseball are anxious to keep their position. Maybe Sammy Sosa didn't purposefully bring a corked bat into a game in June 2003, as he has claimed. And maybe Sosa doesn't use steroids as is widely rumored. But if he did these things it wouldn't be that surprising: As a slugger like Sosa gets older, it's harder for him to sustain the muscle and power needed to drive balls over the fence and justify his multimillion-dollar salary. It's natural to look for an edge in this kind of situation.

Professional sports are an extreme environment. Success can transform you into a cultural icon and a centamillionaire, while failure can leave you injured, broke, and barely employable. People act in extreme ways with stakes like these. In a survey of 198 top athletes conducted in 1995, more than half indicated they would take a drug that would help them win every competition for a five-year period—*even if they knew that at the end of five years the drug's side effects would kill them.*[20]

Other sports have also seen an intensification of winner-take-all inequities in recent years—and rampant doping by athletes. Professional cycling is one such sport.

Superstar cyclist Lance Armstrong is a hero in the United States. He gives speeches for $200,000 a pop, higher than Bill Clinton's rate. His memoir, *It's Not About the Bike,* spent months on bestseller lists. Armstrong's appeal is obvious: He's a man who defied death and went on to become a better athlete than before—and a multimillionaire. Cancer cost him a testicle and made him sterile, yet he has three children thanks to frozen sperm. He says his illness helped to steel him for the intense punishment of professional cycling and the grueling ordeal of the Tour de France, an annual race that covers 2,000 miles over 23 days. Quite apart from all these superhuman qualities, Armstrong is so admired in the United States because he beats the tights off French riders every year. Any nemesis of France has got to be all right.

Armstrong's reception is very different on the other side of the Atlantic. European journalists have dug in his trash and combed through his past, hunting for evidence of drug use. The French authorities investigated him and his teammates for two years, summoning them for drug testing and questioning.

The animosity toward Armstrong is not as simple as continental pride. For one thing, the European public happens to know quite a bit about professional cycling, unlike Americans, and they know that doping is pervasive among professional cyclists. In 1998, Willy Voet, a key staff member for the top cycling team Festina, was arrested by French customs agents with several coolers full of pharmacological wonders. The following year Voet published a bestseller entitled *Chain Massacre: Revelations of 30 Years of Cheating* that claimed that nearly all top riders were doping. Those who weren't inevitably lagged at the "back of the pack." Voet's revelations largely confirmed conventional wisdom and gave Europeans good reason to be skeptical that a cyclist like Armstrong could not be on drugs when he beats the best cyclists of the world who *are* on drugs.[21]

The Europeans also know that drug testing during the Tour de France is a joke. The most widely abused performance-enhancing drug in cycling is erythropoietin (EPO), an artificial hormone that allows the blood to carry more oxygen, thus boosting endurance. EPO was originally conceived to help fight cancer and kidney problems (Armstrong admits taking it when he was ill), but it soon swept the sports world, especially professional cycling. A huge black market for the drug thrives in both Europe and the United States—a black market fueled by shameful covert appeals to athletes by the drug's licensed makers, including Johnson & Johnson. EPO is a problematic drug for professional sports. Because EPO exits the bloodstream in just a few days, even sophisticated drug tests can't provide evidence that an athlete was using EPO during his training. "Athletes don't tend to stay on EPO year-round," explains Dr. Michael Ashenden, a leading expert on doping in sports. "The athlete typically has to take EPO for three or four weeks to gain the maximum advantage, and once they've raised their blood-cell mass, they can lower it to maintenance level. So, if the athlete is cunning...there is no way to detect it even though they still have the benefits from it weeks later." Also, diluting EPO with a saline solution can make it difficult for blood tests to detect its presence even in an active user. "A racer who gets caught by doping control is as dumb as a mule," said Willy Voet.[22] (Although not nearly as dumb as the staff member of a cycling team who gets busted with a whole cooler of drugs.)

Armstrong says that persistence is his secret formula for success, not an EPO drip bag. "What am I on?" Armstrong asks in a

Nike commercial that ran in France. "I'm on my bike six hours a day. What are you on?"

Whatever the truth about Armstrong, peering into the high-stakes world of professional cycling further helps to illustrate the powerful logic behind cheating in today's winner-take-all sports world.

If you're an aspiring top cyclist, you quickly learn three central facts about your profession. One, many cyclists cheat, especially in Europe. To not dope while competing in Europe is akin to playing by your own rules, rather than the prevailing rules of the sport. "A lot of us were really naïve when we first went to Europe," says Marty Jemison, who raced on the Postal Service team with Armstrong in the late 1990s. What the American riders discovered was that there was "a race within a race" as the European cyclists and their doctors tried to manipulate human physiology, often with newly created designer drugs. "It changes every year. Every year is a learning experience of medicine," says Jemison.

The second fact you learn as a cyclist is that it's very difficult to get caught if you're using EPO and other drugs. Cycling has drug-testing practices that look tough, but actually getting caught is a different matter altogether, given the difficulty of detecting the drugs. Cyclists see both their own personal doctors and the team doctor, says Jemison. "Your team doctor might not know what your doctor gives you. They won't question something if the results are good. They can be fired if a team doesn't perform. So maybe their job is to make sure that someone just doesn't *test* positive."

The third fact you learn is that the rewards of using drugs and winning can be astronomical. Twenty years ago, a top American cyclist would be regarded as an oddity and yet manage to make a decent living. Now cyclists can make a huge amount of money through lucrative sponsorships and endorsements. While Lance Armstrong's annual income of $15 million is an extreme case,[23] a few other top cyclists also rake it in—as long as they consistently win races. No sponsor wants to back a loser. In contrast, merely average cyclists struggle just to keep riding. "There are very few people in the cycling world getting rich from it," says Gerard Bisceglia, who runs USA Cycling, the leading association of cyclists. "The top rider makes more than the next five combined."

Professional cycling is a murderous sport, and it's not uncommon for cyclists to ride 20,000 miles a year as part of their training. For all professional riders, but especially those at the bottom, the sense of insecurity and pressure is unrelenting. "You know you have to ride well, that's why you train so much," says Frankie Andreu, who has raced with Armstrong in Europe.

Many aspiring pro cyclists don't have a whole lot to fall back on, as is typical of other athletes. Their intense years of training may have pushed aside any serious college education, and they may be essentially without job skills.

Marty Jemison, who has heard the din of the cheering crowds on the Tour de France, never made it to the top ranks of cyclists. These days he runs a small bike-tour group in Utah. "When the screws are turned so tight, it's like war," Jemison says of cycling. "It's survival of the fittest, and if people don't find their way, they're out. It's so competitive." Jemison says that he did not use drugs himself. "It was a joke among the French about me, that I could have been great but that I 'rode with water,'" which means riding clean.

The French riders were right, Jemison thinks. "I could have been great, and financially it could have changed my life and made it a lot easier for me now, but you make your own choices. You're only responsible for yourself."

The winner-take-all phenomenon seen in sports is found everywhere. The pay gap between top heart surgeons and typical general practitioners is huge and much greater now than it was thirty years ago, as is the gap between star corporate law partners and lowly associates, between telegenic stock analysts who hold court on CNBC and the grunts that churn out much of Wall Street's research, between top scientists who strike gold with the right patent and more average lab rats, between pop stars with a global audience and unsigned musicians who tour in cities like Wichita.

Robert Frank, a Cornell economist, first began noticing the winner-take-all phenomenon in the 1980s. It wasn't hard to spot this trend, since as an academic, Frank worked in a star system himself. The very top university professors increasingly commanded six-figure salaries in the 1980s even as the average pay of professors inched up only slightly. Looking at other professions, Frank saw these same trends at work. In 1988, he teamed with an old classmate from his graduate days at Berkeley, Philip Cook, and began researching the new pay gaps. The result of their work, *The Winner-Take-All Society*, published in 1995, quickly became a classic in the field of economics.

Professional cycling provides one example of several of the winner-take-all dynamics that Frank and Cook analyze. Television audiences for the Tour de France have expanded rapidly in the past decade, with more fans watching from beyond Europe via satellite TV. Companies sponsoring a top rider like Lance Armstrong—in effect, plastering him with logos—can thus expect to have their brand identity beamed to more people worldwide. Such exposure is more valuable now than it once was, since many companies are global in scope and are competing for market share in many more places than they were a decade ago. An increasingly competitive economy also vastly reinforces the commercial value of someone like Armstrong who achieves a well-known "brand identity" status. If you get that athlete to attach his brand identity to your product, your product has a much better chance of standing out against the noise of an ever-more-crowded advertising marketplace. And that is worth a huge amount of money.

In turn, once the value of an athlete's association is recognized by one company which pays handsomely for that association, this value can feed on itself through "self-reinforcing processes," as Frank and Cook write. Everyone wants what is hot, and the clamor further increases the value of whatever becomes hot. Through this upward spiral, top performers can create an unbelievable earnings gap between themselves and other performers, even ones just as talented. "[A] small initial advantage can eventually engender a nearly insurmountable lead," note Frank and Cook.[24]

Lance Armstrong's relationship with the U.S. Postal Service shows just how high the value of a single superstar can be in comparison to others. Following his third victory in the 2001 Tour de France, Armstrong renegotiated his sponsorship with the U.S. Postal Service and landed a deal worth $4 million a year. The Postal Service justified this as a savvy marketing move for an institution that's fighting for its life against UPS, FedEx, and e-mail. "Lance is about perseverance, and so is the Postal Service," said a spokeswoman, apparently referring to the old "rain, sleet, or snow" promise.[25]

News of this hefty payout was not greeted warmly among many postal employees. How was it, they wondered, that the Postal Service could pay millions of dollars a year to a *bicyclist* when many workers who had dedicated their careers to the Postal Service struggled to get by on mediocre pay? "I felt sick to my stomach reading that figure," one mail sorter said after the deal.[26] The average postal worker earns about $39,000 a year, which means that Armstrong is making as much money annually as 102 postal workers combined—workers who cumulatively put in over 200,000 hours a year.

What does Armstrong do for a payout that amounts to over $11,000 a day? Mainly, his labors consist of one simple act. Whenever he dresses for a race—or, as important, a photo shoot—he pulls on a biking jersey that features the Postal Service logo on both the front and back.

Pretty good work if you can get it.

WRITERS LIVE ON an entirely different planet than professional athletes. But here, too, winner-take-all dynamics have reshaped the field and are linked to a rise in cheating.

Jayson Blair is now famous as the lying journalist who triggered an uproar at the *New York Times* and brought about the fall of its imperious editor, Howell Raines. The story was remarkable in some ways—never had the *Times* been turned so upside down. But it was otherwise familiar. Hadn't something like this just happened a few years earlier?

In May 1998, Adam Penenberg, an editor at the online magazine *Forbes Digital Tool,* found himself frustrated by his efforts to further investigate an incident he had read about in the *New Republic,* where a fifteen-year-old Bethesda computer hacker had penetrated the security of a company named Jukt Micronics. Penenberg couldn't find any record of the company's existence. So he called the writer of the story and asked for more information.

Stephen Glass, a twenty-five-year-old hotshot at the *New Republic,* seemed happy to help Penenberg out. He gave him the phone number of the company, and also directed him to its Web site on AOL. Penenberg eventually realized that both were fakes. The number was the cell phone of Glass's brother; the Web site had been created by Glass himself. Jukt Micronics didn't exist; nor, needless to say, did the fifteen-year-old hacker. A full-blown meltdown ensued at the *New Republic* as Glass's long history of deceptions came to light.

Glass has since gone on to more lucrative pursuits. He went to Georgetown Law School and nailed a marketable degree. And his journalistic dishonesty was lavishly rewarded by Simon and Schuster, which paid him a six-figure advance to write an autobiographical novel. The timing of Glass's novel was fortuitous. In May 2003, as Glass was making the rounds on *60 Minutes* and elsewhere to promote his book, Jayson Blair was fired from the *New York Times* for a pattern of behavior very similar to Glass's activities at the *New Republic*.

The Glass and Blair episodes each triggered a huge amount of media attention. But most of the analysis in the wake of both scandals failed to offer any compelling answers as to why promising young journalists would possibly fabricate or plagiarize material on a large scale. The only explanation to many observers in the Glass case was that the guy must have suffered from some kind of sociopathology, a view that Glass gives some credence to in his novel, which features a protagonist with an excessive desire to be loved. A more complicated set of motives has been imputed to Jayson Blair. These include not just psychological problems but drug addiction and alcoholism as well. Blair has said that all of these may have played a role. It's also been alleged that Blair got away with cheating because he was black and was coddled by a newspaper mindlessly committed to affirmative action. (An explanation that Blair has said is absurd.)

Only a few postmortems of these journalistic scandals have focused on the most obvious possible motive for why young, ambitious professionals might take such big risks—to reap big rewards.

Fabrications by journalists are nothing new nor are conflicts of interest in the media. But while there is no hard evidence that misconduct in journalism has increased in recent years, there are plenty of reasons to think that journalists are facing new pressures on their integrity that stem from a greater focus on the bottom line and bigger pay disparities.

Most people are familiar with the ways in which a growing obsession with profits has undermined the media's watchdog role and propelled it toward a crasser, less credible focus on "infotainment." Public trust in the media has fallen, and many journalists share the public's uneasiness with their profession. According to a study by the Pew Center on the People and the Press, "Majorities of working journalists say that increased bottom-line pressure is hurting the quality of coverage. This view is more common than it was just four years ago." According to another survey, 63 percent of journalists perceive a decline of ethics and values in the profession.[27]

Less familiar—and more relevant to the recent spate of plagiarism cases—are the ways in which many journalists are scrambling to score financially. "Everyone in journalism wants to make as much money as the lawyers and various other people they write about," commented Richard Blow, the former Washington editor of *George* magazine, who had briefly employed Stephen Glass.[28] A generation ago, making big money wasn't a realistic option for most journalists. Now it's commonplace—*if* you can become a star.

In the world of opinion journalism that Glass inhabited, the path to stardom is to write hit pieces that draw attention and heat, ideally at the expense of people more powerful than yourself. Then come the calls from the talk-show producers and lecture agents and book publishers. If you're lucky, you can end up like Tucker Carlson, the bow-tied *Weekly Standard* prodigy who now is a co-host of *Crossfire* and commands big bucks for his various activities; or David Brock, an opinion writer of few scruples who turned himself into a millionaire by attacking Anita Hill and the Clintons, and bought his own townhouse in Washington before the age of thirty; or Michael Lewis, another opinion writer who scored early with his first book and has been commanding fat six-figure advances ever since. To increase one's odds of making it big, it helps to write a lot of stories at a fast clip—to build up the kind of cachet and brand identity that can be parlayed into high earnings. Stephen Glass was in such a hurry to get big that he overcommitted himself, taking on assignments for *Harper's*, *George*, and *Rolling Stone* even as he tried to deliver on all his responsibilities at the *New Republic*.

The path to stardom for newspaper reporters is different, but can be even more lucrative. A reporter with large ambitions tries to get assigned to the big stories of the moment and then do well enough to get noticed and have a wider choice of beats. Eventually he can find a hot story that can be parlayed into a book deal. Blair tried to do this unsuccessfully with the Maryland sniper story, shopping around a book about the case. Many other journalists have done it with far greater success: Thomas Friedman's book on the Middle East, *From Beirut to Jerusalem*, catapulted him into the ranks of America's most successful journalists and launched him on the road to being a millionaire. *Newsweek*'s Michael Isakoff stumbled on the hot story of a lifetime when he became closely involved in the Clinton sex scandal—and cashed in his experiences for a huge book advance. Joe Klein's coverage of Bill Clinton's presidential campaign gave him the insights he needed to write *Primary Colors*—and rake in a fortune as a bestselling author.

Thirty years ago Bob Woodward and Carl Bernstein became very wealthy young men when their *Washington Post* articles on Watergate helped bring down President Nixon. Their sudden transformation was something of an oddity, and other journalists marveled at their riches and newly lavish lifestyles. Today this kind of thing is far more common. "Journalism didn't used to appeal to people who wanted to become famous," explained Charles Peters, editor of the *Washington Monthly*. "Now you've got people drawn to Washington who used to be drawn exclusively to New York or L.A.—Washington journalism has become another path to becoming famous."[29]

Yet for all the new examples of journalists striking it rich, the financial situation for most journalists has actually grown more precarious over the past two decades. Writing opinion commentary has always been a dubious career choice, but it is even more so today given the high cost of living in places like Washington, D.C., and New York City. A few decades ago, an intellectual with a clever pen could take solace in low housing prices and manage to make ends meet. For example, in the 1960s my father bought a six-bedroom house with a river view in a prime New York suburb on the salary he earned as an associate editor of *Commonweal*, a small weekly opinion magazine. Now that same house is out of reach even for a young corporate lawyer. The only opinion journalists who live well in today's America are among the lucky few who either have "broken out" or have a source of income unrelated to their job, like a trust fund or a rich spouse.

Even journalists at top news organizations often are barely able to afford to live in the cities that they cover. For example, a reporter at *Time* magazine with seven years' experience can expect to make roughly $75,000, while a reporter who's been at the *Wall Street Journal* for four years is probably pulling down around $55,000. This kind of money doesn't go very far in New York City. I know one journalist who moved to New York to take a plum job in television news, only to find that she couldn't afford an apartment on her salary. Lucky for her, she managed to bribe a building superintendent to get a rent-stabilized place. Reporters at other top publications like the *Washington Post*, the *Los Angeles Times*, the *Boston Globe*, and the *San Francisco Chronicle* are paid well compared to reporters working for smaller papers, who have a median salary of $40,000 a year. Yet all of the big cities have become a lot less affordable over the past decade.[30] At the same time, many more young journalists are starting their careers with large loads of student debt incurred by master's programs in journalism—an expensive credentialing process that's become more common as competition has increased for the better jobs in journalism.

In short, journalism has become yet another winner-take-all arena over the past two decades. So has the realm of book publishing, where name-brand authors can make millions of dollars writing a book every year or two—while merely ordinary authors find publishers less willing to invest in the careers of "mid-list" writers. Not surprisingly perhaps, book publishing, too, has been beset by an unprecedented number of plagiarism cases in recent years.

Winner-take-all trends may not be the only explanation for the cheating of Jayson Blair and Stephen Glass, or the recent ethical problems of other journalists over the past decade, including Mike Barnicle, Michael Finkel, Ruth Shalit, Rick Bragg, Patricia Smith, Monica Crowley, Mike Hornung, and a dozen others.[31] Nor may these trends account for the spate of scandals involving book authors in recent years, including Stephen Ambrose, Doris Kearns Goodwin, Michael Bellesiles, and Brian VanDeMark. In each of these cases, there are different possible explanations for wrongdoing: a psychological breakdown, a sloppy research assistant, an erroneous pasting of words from a Nexis file, a political ax to grind. It also may be that the flurry of so many scandals involving ambitious journalists and authors in recent years is pure coincidence, that it has nothing to do with the bigger rewards now being dangled before writers along with the greater economic insecurities they face.

Maybe. But I suspect otherwise.

It's intuitive that as the rewards at the top become bigger, more people will do anything to get to the top. What is less easy to understand is the pressure and anxiety felt by those people who, by most measures, are doing very well already. The precepts of laissez-faire ideology suggest that inequality is okay on principle, but especially okay if everyone is getting richer. Yet in the real world, big pay gaps can have a corrosive effect on the integrity of even those people who should be extremely grateful for all the money they are earning.

Consider a rookie ballplayer. He earns $300,000 a year, mainly for sitting on the bench. He is a legend back in his hometown and should consider himself a lucky man. Unfortunately, when he compares himself to other people, he's not thinking about the guys he played ball with in high school who are now auto mechanics and accountants. He's more likely to look up than down—comparing himself to the multimillionaire celebrities he works with every day. As he does, he may not feel so lucky. He might even feel poor.

The reason for this is that most human beings think about their well-being in terms *relative* to those who share their immediate community, as Thorstein Veblen pointed out a century ago in *The Theory of the Leisure Class* and as Robert Frank has discussed in some detail in his book *Luxury Fever.* Absolute well-being doesn't matter as much as it should. Most of us would rather earn $100,000 a year in an organization where nobody makes more than $90,000 than make $110,000 at a job where all our colleagues are paid $200,000. We'd feel better about ourselves if we owned a '97 Toyota Camry in a neighborhood where everyone else drives '88 Honda Civics and '90 Mazda Protégés, than if we owned a brand-new Camry in a neighborhood filled with Jaguars and Mercedes.[32]

The notion that people worry more about their place in the economic pecking order rather than the size of their paycheck has found support in research exploring the interplay of money, hierarchy, and happiness. Studies by biologists and health researchers also suggest that being in a subordinate position can do a hatchet job on your self-esteem, leave you chronically stressed out, and undermine your physical health. A famous long-term study of thousands of British civil servants found that lower-ranked employees died earlier—even when researchers controlled for diet and personal habits like smoking. Stress and "low job control" appeared to explain the difference in mortality rates.[33]

Concerns about relative position are not simply the product of envy and other shallow emotions. We compare ourselves to others for very good reasons. If you're wearing a $500 suit and another job applicant sports a $1,000 suit, both of you are wearing nice suits. But the other guy may have an advantage, all other things being equal. If you're living on a street where everyone has 10,000-square-foot mansions and your home is 6,000 square feet, you and your neighbors all have plenty of room. But when you throw a party and invite lots of professional acquaintances, they'll see that you're the poor person on the block. Maybe they'll be less likely to think you're a rising star after all, and won't throw venture capital, or big contracts, or whatever, your way. And God help the would-be mogul who shows up in Jackson Hole on a *commercial* flight.

Frank and others argue that, ultimately, anxieties about relative position reflect evolutionary imperatives shaped by a long human history in which small advantages over others translated into a better chance to survive and reproduce. "There is compelling evidence that concern about relative position is a deep-rooted and ineradicable element of human nature," Frank writes.[34] So go ahead, feel sorry for that baseball benchwarmer pulling in paychecks bigger than anything most of us will see in our lifetime—but who shares a locker room with guys who make $10 million a year.

Feel sorrier, though, for the sales manager at a Banana Republic who can barely make ends meet at a job selling expensive clothes to young professionals who make five times what she does. Worries about relative position are most wrenching when people are hurting economically and when competitive emotions are mixed with survival instincts. This is exactly the situation for tens of millions of Americans who were bypassed by the boom—yet see its fruits displayed before them every day.

THE FALL OF TRUST in the United States over the past forty years has long been discussed and debated. It is well known that Americans trust nearly every institution less than we used to. We're less trusting of government, less trusting of the media, less trusting of religious institutions, and less trusting of lawyers and other professionals.

The falling trust in various professions is especially notable. Americans are more fearful of being ripped off, misled, or otherwise cheated by people who are charging us money for services or whom we are relying upon to advise us on key parts of our lives.[35]

Americans have also become less trusting of each other. In 1960, 58 percent of Americans agreed that "most people can be trusted." By 1998, only 40 percent agreed with this statement. Every few years for the past quarter century, the General Social Survey (GSS) has asked hundreds of Americans a telling question about trust: "Do you think that most people would try to take advantage of you if they got the chance, or would they try to be fair?" When the GSS first started asking this question in the 1970s a large majority of Americans didn't fear being cheated. But such fears increased during the 1980s, and by the late 1990s, nearly as many Americans thought that most people would try to take advantage as thought that most people would try to be fair. Sixty percent of Americans now say that "you can't be too careful in dealing with people."[36]

Distrust is obvious fuel for cheating. If you think people are out to cheat you, you're more apt to believe that rules don't really matter and that you've got to live by your wits as opposed to ethical principles. You may imagine for self-protective reasons that you need to cheat others before they get a chance to cheat you.

For all the talk over many years of rising distrust, only recently have scholars begun to make the link between inequality and distrust. The notion of such a link rests, in part, on common sense. If you don't see yourself as doing well economically in relative terms and if you think the system is stacked against you, it's easy to be pessimistic and resentful. In contrast, feelings of trust are associated with optimism about the future and goodwill toward others. In his book *The Moral Foundations of Trust,* scholar Eric Uslaner used a variety of opinion surveys taken over the past several decades to examine how and why people trust others. He writes: "If you believe that things are going to get better—and that you have the capacity to control your life—trusting others isn't so risky. Generalized trusters are happier in their personal lives and believe they are masters of their own fate."[37]

It's not easy to feel like you can control your life in America's postindustrial economy. In a winner-take-all market plagued by stagnating wages, downsizing, and rising prices for key life necessities like health care and housing, many people have good reasons to be pessimistic and resentful. Polling during the boom periods of both the 1980s and 1990s showed that even as the economy grew by leaps and bounds, many people didn't believe that their own incomes would rise. More than half of Americans consistently said that they weren't making enough money to lead the life they wanted, and many didn't see such money in their future. Instead, large percentages of Americans worried that their own financial situation might well deteriorate and also worried about their children's prospects. Upward of half of Americans, for example, felt that their children would be worse off than they were.[38]

The prevalence of such views might be understandable during a prolonged recession. But it's remarkable that so many Americans would feel this way even as the nation as a whole grew wealthier than ever before in history. The vitality of the U.S. economy was among the most celebrated aspects of American life during the late 1990s when the American economic model had conquered the world, when kids in their twenties could make millions of dollars, when unemployment had fallen to a thirty-year low, and when some observers were predicting an end to the business cycle, with its booms and busts. Self-congratulation was the mood of the moment, as it had been in the mid-1980s, when it was "morning again in America." And yet up to half of Americans during both of these periods felt they weren't earning enough money and their kids would be worse off than they were.

Why did so many people believe this? Because it was true. Most of the gains from the boom went to the top 20 percent of households, while many households lost ground. Being a middle- or lower-income American during the '80s and '90s was akin to sitting through a long and rowdy victory party—when you're from the losing team.

The divisive effects of inequality have been further aggravated by the way in which large income gaps have pulled American society apart culturally and geographically, slicing it up into different groups that have little in common. The size of our paychecks determines where we live, what we wear, what we drive, what beer we drink, what kinds of restaurants we choose, what we watch on television, where we work out, where our children go to school, where we vacation, what hobbies we engage in, and much more. All of these features of our lives help shape if not our own class identity, then certainly the class labels that others put on us. Dramatically uneven levels of income result, inevitably, in big disparities in class identities across a society. Since most people feel more comfortable around those who are having a similar life experience, the divisive potential of these disparities is obvious.

Class is nothing new to Americans, even if we've tended to deny its existence here. But the intensity of class divisions has waxed and waned over two centuries, along with levels of inequality. Class was a powerful aspect of American life in the Gilded Age, and again in the 1920s, two periods that were characterized by many scandals. The middle decades of the twentieth century saw these divisions fade substantially. The 1940s through the 1960s have been called the "Great Compression," because of the dramatic narrowing of income gaps that occurred during this period. During these prosperous decades, all classes of Americans got richer at roughly the same rate. Narrowed income gaps, in turn, produced one of the most socially egalitarian eras in American history (although this was also an era marked by pervasive race and gender discrimination).[39] The nation's imagination was captured by the ideal of a universal middle class that could, and should, encompass everyone. Identical suburban homes stand as an icon of the early postwar period for good reason.

The egalitarian mood of the day was reflected in the executive suites of corporate America, where top executives understood that they would not be granted salaries that too greatly dwarfed those of average workers. Sociologists and economists speak of the "equity norms" that prevailed in business in the early postwar period. In 1965, CEOs made on average fifty times more than the typical worker. While large, this gap is nothing compared to today, when CEOs make nearly 300 times what the average worker makes. Before the 1980s, pay gaps were still small enough that many CEOs didn't imagine themselves as some separate imperial breed of leaders. For example, members of the Harvard Business School Class of 1949, a third of whom became CEOs, exemplified the everyman sensibility of yesterday's corporate leaders. Most of them lived in comfortable but modest homes and frowned on conspicuous consumption. They drove average-priced cars, and saw themselves as lucky to have the opportunities that they did.[40]

Not surprisingly, the early decades of the postwar period were a time of enormous social solidarity and trust. The early postwar years were also an era of comparatively little cheating in business and other sectors of American society, which makes sense. The middle class perceived that the social contract was delivering on its promise and felt respectfully treated by those higher on the economic ladder. The rich, in turn, were not living radically different lives than the middle class and weren't able to spin off easily into a separate moral reality governed by its own rules.

That was then. As the income differences among Americans have grown larger in recent decades, so have social differences. The enduring correlation between ethnicity and income aggravates the problem, piling ethnic and cultural differences on top of class differences. Looking at each other across the chasms of class and race, many Americans see little reason to believe that they share each other's values—and little reason to trust each other. "Trust cannot thrive in an unequal world," writes Uslaner. "People at the top will have no reason to trust those below them.... And those at the bottom have little reason to believe that they will get a fair shake."[41]

In the past decade, geographic divisions among Americans by income have become especially noticeable. Many working-class people find themselves priced out of the areas where they grew up, yet their services are still needed in these communities. And so you see car mechanics, garbage collectors, and police officers driving long distances every day to work in towns or cities that used to be affordable to blue-collar people. This kind of thing—residential segregation by income—has deepened as inequality has grown in the past quarter of a century.[42] Much residential segregation is explained by the simple fact that people live where they can afford to rent or buy a place. But more deliberate self-segregation by the affluent is on the rise. Wealthy enclaves have always existed—places like Beverly Hills and Palm Beach and Sutton Place—but they tended be small in size and unusual. Now, homogenous communities of affluent and semi-affluent Americans are both more numerous and larger in scope. In the early 1970s, there were roughly 2,000 private "gated communities" in America where access was restricted to members and visitors. Today, there are more than 50,000 such communities. Some seven million households now live in gated communities, and 40 percent of new homes built in California are in gated communities. "What is the measure of nationhood when the divisions between neighborhoods require guards and fences to keep out other citizens?" ask Edward Blakely and Mary Gail Snyder in their book, *Fortress America*. "Can the nation fulfill its social contract in the absence of social contact?" Blakely and Snyder, along with many other social observers, answer an emphatic no to this question.[43]

When I first started investigating cheating, my guiding assumption was that nobody *wants* to cheat. I still think that. No athlete wants to pump his body full of drugs that shrink his testicles, or change the shape of his head, or could turn his blood into molasses and leave him dead halfway through the Tour de France. No stock analyst wants to go on CNBC and hype a stock that every insider knows is a piece of junk. No chief financial officer wants to cook earnings reports, and no accountant wants to rubber-stamp these reports. No journalist wants to make up her sources.

But when you look at the effects of inequality in our society, you can understand why respectable people consistently do all of these things. The winner-take-all economy has loaded up the rewards for those who make it into the Winning Class, and has left everyone else with little security and lots of anxiety. Inequality has also pulled us apart, weakening our faith that others follow the same rules that we do.

Unfortunately it gets worse. Two decades of change in American economic life—and a steady string of victories for laissez-faire ideologues—hasn't just shifted the financial incentives for individuals or the operating strategies of business organizations. It has deeply affected American culture overall, reshaping nearly everyone's values.

And not for the better.

CHAPTER FOUR

A Question of Character

WHY DID SCOTT SULLIVAN TURN BAD?

A few years ago, Sullivan was among the most respected young corporate leaders in America. He was the princely chief financial officer of the telecom giant WorldCom and the confidante of its CEO, Bernard Ebbers. Sullivan had come to WorldCom in 1992, at the age of thirty-one, when it was still a modest-sized company. Known for his brilliance and intense work routine, he rose quickly in the ranks. By his mid-thirties he was a top figure in one of America's fastest-growing corporations. Many analysts saw Sullivan as the brains behind Ebbers, the smarter half of what Wall Street called the "Scott 'n' Bernie show." It was Sullivan who orchestrated WorldCom's acquisition of MCI in 1998. It was Sullivan who worked the investment world, finding the cash to fuel WorldCom's extraordinary growth.

Ebbers was two decades older than Sullivan and towered over him by a foot, but the joke was that the two men finished each other's sentences. In nearly daily private lunches with the CEO, Sullivan helped map out the business strategy that involved a steady stream of mergers and deals. These aggressive tactics turned WorldCom into one of America's biggest companies and boosted its stock by 7,000 percent during the 1990s.[1]

In 1998, *CFO* magazine bestowed Sullivan with the annual CFO Excellence Award. He was thirty-seven and making $20 million a year. The award came with a gala fete in Palm Beach, and Ebbers made a surprise appearance to toast WorldCom's rising star. "I don't think WorldCom would be where it is today without Scott," Ebbers said.

Sullivan lived the good life with his new money and status. He and his wife, Carla, were VIP guests at the Super Bowl and the Stanley Cup and the Olympics. He was invited into high-society circles in Boca Raton, where he owned a home, and he began laying the foundation for a much larger presence in this rarefied world by starting construction on a 16,000-square-foot Mediterranean-style mansion in the tony Boca Raton community of Le Lac.

The house would have suited a Roman proconsul. Plans called for eight master bedrooms, nearly all with a Jacuzzi bath. The gigantic master suite featured his and hers showers and toilets, as well as a minipool. There was to be an eighteen-seat movie theater, a storage room just for furs, an elaborate wine cellar and tasting room, a six-car garage, a two-story boathouse with maid's quarters, and 7,000 square feet of terraces and balconies and walkways. Costs for constructing the mansion and its four outlying buildings were estimated at $15 million.

The old guard at Boca Raton might not be impressed by Scott Sullivan's lineage, which traced to a middle-class family in upstate New York, but surely they'd be impressed with his beautiful home. And appropriately, the lavishness of Sullivan's real estate ambitions could not possibly overshadow those of his boss. Ebbers had earlier shelled out $47 million for a 500,000-acre working ranch in British Columbia that included 20,000 head of cattle and a luxury fishing village. He also had a mammoth yacht named *Acquasition.*

Sullivan's mansion is not yet finished. He was fired from WorldCom on June 25, 2002, amid allegations of massive fraud. And on a warm morning a few weeks later, he entered the Manhattan federal courthouse wearing handcuffs. Sullivan was flanked by FBI agents, each gripping an arm—as if he might try to make a dash for freedom at any moment. A sea of cameras recorded Sullivan's "perp walk," and a crowd cheered at the spectacle of the disgraced corporate prince. The scene had been carefully staged by prosecutors anxious to show that they were cleaning up big business. "No more easy money for corporate criminals," President George W. Bush said two days before Sullivan's perp walk, as Bush signed a new law governing corporate accounting procedures. "Just hard time." Inside the courthouse, Sullivan was arraigned on seven felony charges, including securities fraud and conspiracy.

It was easy for Americans to hate Sullivan that August day. Enron's demise the previous fall had been followed by months of shocking revelations about criminal greed in the executive suites of many corporations. The stock market was in a downward spiral, with the Dow dropping nearly 3,000. A stunning $7 trillion in investor wealth had simply vanished between early 2000 and the day that Scott Sullivan was arraigned in New York. Nearly a third of this money had been in telecom stocks. In some cases, investors had only themselves to blame for their losses. Anxious to get rich quick, investors traded high-tech stocks like weekend gamblers in Las Vegas and paid the price when the market went south. But the more typical victims of the crash were millions of Americans who deferred to the judgment of their brokers or whose retirement savings were in large pension funds. These investors felt deeply betrayed as they watched their portfolios shrink. They hated Scott Sullivan and his kind for good reason.

Yet Sullivan was an unlikely corporate villain. He grew up in modest surroundings in Bethlehem, New York, and attended public high school. From there it was on to Oswego State University, where he majored in business and accounting, graduating summa cum laude. One of his professors later remembered him as a mature, straight-arrow type who had "a unique ability to get along with others." He was offered jobs with six top accounting firms, and chose to stay close to home with a position at KPMG in Albany. He did extraordinarily well in that job, and moved on to a telecommunications company in south Florida, which later merged with WorldCom. Sullivan caught the eye of Bernard Ebbers with his strategic savvy and his incredible hard work—colleagues would find voice mail from him at two or three in the morning. Soon he was one of "Bernie's boys." Sullivan was named CFO of WorldCom in 1994, at the age of thirty-three. It was a remarkable rise.

WorldCom's steel-and-glass headquarters towers over everything else in Clinton, Mississippi. Before its fall, WorldCom was the only *Fortune* 500 company in the state, and its top executives were treated as local heroes. Sullivan was among the most respected people in the company, known for taking care of his staff. Sullivan even shared some of his 2000 bonus money with at least seven members of his team, writing each of them a personal check for $10,000. His reputation was also luminous in the industry at large. "He was one of us, an analyst's CFO. He knew the numbers," said Patrick Comack, a telecom analyst. Compared to Ebbers, Sullivan was always seen as mild mannered and was respected as a sober voice of reason in a telecom industry filled with wild men. All in all, he seemed to be one of the good guys in corporate America. And back at Oswego State University, Sullivan was a legend—an upstate boy who had conquered the business world and remembered the little people, actively contributing both time and money to improve his alma mater.[2]

Those who knew Sullivan well also knew that his life was not all it seemed on the outside. Yes, Sullivan was a confidante of Ebbers and had unparalleled access to WorldCom's CEO, but he told friends that he loathed the brash Ebbers. Yes, Sullivan and his wife circulated in high society, but their private life was filled with challenges. Carla was in constant poor health, and because neither of them liked Mississippi, Carla stayed in Boca Raton with their adopted daughter while Sullivan commuted back and forth to Clinton. As for the mansion, this kind of conspicuous consumption was a new thing for Sullivan. For a decade he and Carla had lived in a modest home in Boca Raton that they had bought in 1990 for $170,000. Sullivan enjoyed deep-sea fishing, but he had never bought anything bigger than an eight-foot boat. Only belatedly, after he'd made millions of dollars and endured years of a killer work routine, had Sullivan begun to live a life of luxury.

When WorldCom's stock was at its height, in early 2000, Scott Sullivan's stake in the company rose above $150 million. He could have cashed out as an immensely wealthy man. Sullivan chose to stay on. And, over the next two years, WorldCom cooked its books with astonishing brazenness while Sullivan was CFO, the corporate official most directly responsible for the veracity of the company's numbers. WorldCom inflated its earnings by some $11 billion through a variety of financial manipulations. When the true accounting was completed by investigators, they alleged that Sullivan had directed the largest corporate fraud in history. His apparent motive was to keep WorldCom and its stock price afloat as the telecom industry went into a nosedive and WorldCom's $40 billion debt became unmanageable. Sullivan's main solution to this crisis was to list hundreds of millions of dollars, eventually billions, of day-to-day operating costs as capital expenses—that is, expenses that fell into the category of long-term investments to improve the company. Capital expenses can be stretched out over many years and Sullivan's switch meant that WorldCom's quarterly earnings appeared to be higher than they were. It was a gigantic lie.[3]

Joseph Wells, a former FBI agent and founder and chairman of the Association of Certified Fraud Examiners, comments that a hallmark of high-level fraud is "rationalization, the ability to call the fraud by a nice name." One way to do this is by reinterpreting accounting rules to give a struggling company more breathing space. Top company officials who engage in fraud say: "I am doing this for the good of everybody who works in the company. I'm not really stealing; I'm borrowing."

Perhaps this is what Sullivan said to himself. Maybe he saw himself as the hero who was keeping WorldCom in business and protecting the jobs of its thousands of employees. Yet Sullivan would have had strong personal motives for fraud as well, since his huge pay packages were pegged to the value of WorldCom's stock, which hinged in turn on company earnings as they were publicly reported. To keep the truth about WorldCom's earnings under the lid, and to keep his own net worth going up, Sullivan lied to his staff and co-workers, to investors, to the media, and, implicitly, to his friends and family.

WorldCom's meltdown dwarfed Enron's in nearly every way. The frauds were much larger, many more people lost their jobs, and the loss to investors was also greater. Over $175 billion in equity value, three times what was lost in the Enron bankruptcy, disappeared as WorldCom disintegrated. Union pension funds lost an estimated $70 billion on WorldCom stock. Major public pension funds in California, New York, and Texas each lost hundreds of millions of dollars, and corporate pension funds and myriad business 401(k)s also took a major hit. These funds had invested so heavily in WorldCom because they believed the company's earnings reports. They also believed leading telecom analyst Jack Grubman, who kept his "buy" recommendation for WorldCom stock almost to the end—and whose huge compensation packages at Salomon Smith Barney were made possible, in part, by the large fees that WorldCom paid that firm.

Many victims will pay the price of WorldCom's crimes over decades. They include workers who will be forced to delay retirement and retirees who must now make do on a smaller fixed income. Unwittingly, these Americans—and the 17,000 WorldCom employees who lost their jobs—found themselves near ground zero of one of the greatest corporate implosions of all time. "WorldCom is the single biggest scandal of them all," comments James Glassman of the American Enterprise Institute. "WorldCom is it."

Scott Sullivan could have prevented this disaster and the incalculable financial pain that resulted. Instead, say prosecutors, he helped orchestrate it.

MAYBE SCOTT SULLIVAN's prison psychiatrist—if he gets one at the Club Fed where he lands—will be able to answer the question of why he turned bad. The rest of us can only offer theories.

I believe that the best explanations of why we have more Scott Sullivans and Barry Bondses and Jayson Blairs boil down to changes in the economy and the new rules that govern it. Our winner-take-all system dangles immense rewards in front of people, bigger than ever before. And today's business culture demands, and glorifies, extreme levels of competitiveness. Meanwhile, the Winning Class has worked to emasculate government regulators in key areas, reshaping the rules so that it can get away with economic murder.

All of this helps to explain a guy like Sullivan. But it's not enough. Human beings are not simply creatures of their economic and legal environment. We don't decide whether to cut corners based only on a rational calculus about potential gains and losses. We filter these decisions through our value systems. And while more of us will do wrong in a system where cheating is normalized or necessary for survival or hugely profitable, some of us will insist on acting with integrity even if doing so runs counter to our self-interest. This is one trait that distinguishes flesh-and-blood Homo sapiens from that consistently rational actor of academic theory, Homo economicus.

In his book *Integrity*, Yale law professor Stephen Carter suggests that integrity requires three steps: discerning what is right and wrong, acting on what you have discerned, and saying openly that you are acting on your understanding of right and wrong.[4] Of course, to show integrity you need to know the difference between right and wrong—which is easier said than done nowadays.

Notions of right and wrong are not only shaped by our family and friends and by work or academic environments but also by the broader culture. As we go through life, the culture's prevailing values, or social norms, shape our ideas about what constitutes the good life, how hard we should labor and to what end, how we should dress and groom ourselves, and much more. Some of us may emerge in early adulthood with a sophisticated ethical outlook picked up from religious education, or from those rare parents and teachers who clearly articulate ideas about character and ethics. "Principled conscience," is how Lawrence Kohlberg, the preeminent theorist of moral development, described the most advanced type of ethical reasoning. Alas, he concluded that the majority of people never get to this stage and tend to have ideas about right and wrong aimed at winning the approval of others and keeping themselves out of trouble. These sorts of considerations don't amount to much of a moral backbone. If you lack principled conscience, your ethics may easily change as the values of the culture change or as you are exposed to different parts of the culture.[5] If Scott Sullivan had stayed in Albany, if there'd never been the crazy boom of the '90s, if he'd never met a shark like Ebbers, if he'd never had a shot at making $50 million a year, if he'd never gotten a taste of Boca Raton's high society, things would have turned out differently for him.

The values of a culture are heavily shaped by the large forces transforming society: war or peace, booms and recessions, demographic shifts and technological change. Values can also be shaped by social movements, religious awakenings, intellectual activism, and celebrity-driven fads—by "influentials" with loud bullhorns who preach a particular way of life. Mass media has made it easier than ever for the values of a society to change quickly.

To understand the ethics of Scott Sullivan and his ilk, you have to understand how the values of American society have changed over the past quarter of a century. Simply put, we have a nastier, more cutthroat set of values than previous generations did. As the race for money and status has intensified, it has become more acceptable for individuals to act opportunistically and dishonestly to get ahead. Notions of integrity have weakened. More of us are willing to make the wrong choices, at least when it comes to money and career.

These are strong statements, I know. It is one thing to say that lawyers, doctors, athletes, and accountants have been corrupted by new workplace pressures. It is quite another to suggest that our entire culture is in deep trouble. Generalizing about the values of 280 million Americans is risky. Pollsters and market researchers do nothing all day but slice and dice the American public into unique segments, of which there are many. The pollster Daniel Yankelovich, for example, identifies eleven core values that capture the allegiance of Americans with varying levels of intensity.[6] Still, there has been a clear shift in our dominant values over recent decades, a shift that reflects the growing influence of market ideology in our society. I see three changes as especially connected to the rise in cheating: individualism has morphed into a harder-edged selfishness; money has become more important to people; and harsher norms of competition have spread, while compassion for the weaker or less capable has waned.

A FEW YEARS BACK, the U.S. Army had a problem. Nobody wanted to join it. Well, not nobody: a steady stream of young recruits signed up, but not enough to meet the army's recruiting goals. Apparently the army's famous slogan, "Be all that you can be," was not capturing the imagination of young people. That slogan was first unveiled in the early 1980s, designed to present the army as a place where members of the "me generation" could realize their potential.

At first the slogan worked. But the boom of the 1990s spelled the end for "Be all that you can be." Labor markets tightened dramatically, and those workers at the bottom of the economic ladder had more job choices. The real problem, though, was the rise of a new kind of radical individualism. Even "Be all that you can be" was too authoritarian for the times. Young people perceived the army as a place where they'd have to follow a lot of rules and would lose their individual identity, so they didn't join.

The army commissioned dense studies of young people by the Rand Corporation, the California think tank best known for its work in the 1960s about how the U.S. could "prevail" in a thermonuclear exchange. The army also hired Dan Yankelovich's polling firm, as well as pricey consultants from McKinsey & Company. The findings from all this research convinced the army that it needed to dispense with "Be all that you can be" and develop a much more hard-edged appeal to individualism. It was not an easy task. "The problem is, how do you attract people who want to be free agents?" asked J. Walker Smith of Yankelovich Partners.[7] We're talking about the army here, not Amway. The military is the total antithesis of individualism. Just look at the buzz cuts, dog tags, green uniforms, and one-size-fits-all body bags.

To solve this marketing riddle, the army hired a big advertising firm in Chicago, Leo Burnett USA. After digesting all the troubling research about the radical individualism of young people, Leo Burnett's ad execs came up with an innovative strategy for suckering these kids into a life of military conformity: lying.

In January 2001, as George W. Bush settled into the White House, a captivating television ad began running on television. A soldier is seen running through the desert, his face set in an expression of steely determination. "I am an army of one," he says. "Even though there are 1,045,690 soldiers just like me, I am my own force.... The might of the U.S. Army doesn't lie in numbers; it lies in me." Two other ads in the new "Army of One" campaign presented young soldiers operating totally alone, calling all their own shots. Somehow Leo Burnett had turned the army into an entrepreneurial experience.[8]

The campaign produced a spike in new recruits, although it did come under some criticism. Bob Garfield from *Advertising Age* made the obvious point: "It's a clever campaign, but substantially dishonest. The Army is not, never has been, and never will be about one soldier." Outgoing Secretary of the Army Louis Caldera defended the campaign. "What we are telling them is that the strength of the Army is in individuals. Yes, you're a member of the team and you've got support from your fellow teammates, but you as an individual make a difference."[9]

If Caldera were honest, he would have told it like it is: that the radical individualism of today's young people makes it extremely difficult to sell any kind of cooperative or communitarian experience. Young people especially don't like rules. It's just the way things are.

Or at least it's the way things have become.

Before the 1960s, individualism in the United States was largely confined to the political sphere.[10] Freedom for individuals meant freedom to speak openly, freedom to worship as we wished, and freedom to live where we wanted. It did not mean freedom to operate outside the norms established by the community, family, and religion. A young man could go west to start his own business and enter politics; a young man could not leave his wife and kids to go west, and he couldn't open an X-rated movie theater once he got there. Among other things, the conformity of the past was shaped by economic hardship. The struggle for basic sustenance dominated the lives of most Americans, and the struggle for physical survival also loomed large before the advent of modern medicine in the mid-twentieth century.

Powerful norms of self-sacrifice shaped people's values in this environment. You existed for your family, and you worked hard to contribute. There wasn't a lot of psychic space left over to focus on your own "needs" and "issues," as today's therapeutic culture uses these terms.

The 1960s famously changed all of that. Self-sacrifice and conformity were rejected and individual self-expression moved to the forefront of American culture, where it has remained ever since. The scholar Ronald Inglehart labeled the new values that emerged from this decade as "post-materialist," in that the triumph over economic adversity that occurred in America and other Western countries during the postwar era allowed people the freedom to turn to other concerns, such as self-expression of various forms—but also to things like environmental protection and animal rights. Daniel Yankelovich, whose research also tracked these changes, labeled this shift of values the "psychology of affluence."[11]

Whatever you call it, the move away from an ethos of sacrifice to one of personal self-interest—to a "new society of individuals"—triggered a social earthquake in Western nations. It was a "psychological reformation as powerful and decisive as the religious reformation of the sixteenth century," according to one team of researchers.[12]

The new individualism was a double-edged sword from the start. It fueled a long overdue assault on social conformity and helped Americans to finally realize the promise of personal freedom. But it also led many people to turn exclusively inward and elevate their own needs above other obligations. The pursuit of shared goals related to community, family, or political life often took a backseat to hedonism, escapism, and endless self-analysis.

Individualism also thrived in a society where old institutions were on the defensive. Organized religion, especially the Catholic church, experienced widespread defections through the '60s and '70s. The traditional family was attacked as a patriarchic tool for oppressing women. Government was discredited by war and scandal. Large corporations were derided as overly conformist and as the enemy of consumers and the environment. Community life in America was undermined by the growing mobility of Americans and the people's urge to escape small-town social strictures.

The rebels of the '60s talked of new humanistic values and institutions that would replace the old order. They imagined a society shaped by the ideals of cooperation and social responsibility, by more enlightened democratic governance, and by sophisticated therapeutic interventions to deal with deviance. It was a nice vision, and parts of it came true. Yet, in the end, many of the old institutions were not replaced by anything, and the new individualism emerged within a social order that did little to counterbalance its worst aspects.

The new individualism quickly came under attack from all sides. On the left, the social critic Christopher Lasch inveighed against an individualistic society run amuck in his 1979 bestseller, *The Culture of Narcissism.* On the right, evangelical Christians spearheaded the most powerful social movement of late-twentieth-century America by tapping into a deep well of anger among traditional Americans who longed for a return to "family values."[13]

None of this slowed individualism's forward march. The arrival of boom times in the 1980s gave fresh energy to the new individualism and morphed it into something very different than what the '60s generation had in mind. It was not a pretty sight.

The "do your own thing" ethos of the '60s rebels unwittingly provided a strong foundation for the laissez-faire revolution of the '80s and '90s. The counterculture, it turned out, emphasized many of the same virtues as free-market ideology, especially individual liberty and choice. It's no surprise that the Libertarian Party was founded in 1972, proclaiming itself the champion of personal freedom through unfettered markets, or that the ideas of Ayn Rand experienced a surge of popularity in the 1970s. Parts of Rand's extreme libertarian philosophy jibed with what a hippie might say. "Man—every man—is an end in himself, not a means to the ends of others," wrote Rand. "He must live for his own sake, neither sacrificing himself to others nor sacrificing others to himself; he must work for his *rational* self-interest, with the achievement of his own happiness as the highest moral purpose of his life." Among the Rand books attracting a fresh audience in the 1970s was her 1961 treatise, *The Virtue of Selfishness.*

While the counterculture of the '60s had targeted consumerism and capitalism in its quest to liberate America's soul, it was not long before many in the counterculture discovered that capitalism could be a potent ally in creating alternate realities within American life. The New Age movement that began in the 1970s exemplified the merging of market and countercultural values. It was a movement, in fact, that drew much of its strength from an aggressive, proselytizing merchant class. Many other Americans with countercultural sympathies also decided that making money was what could really set them free to realize their individualism. As David Brooks has recounted in *Bobos in Paradise,* the '60s ultimately paved the way for a permanent cease-fire in the long war between bohemian and bourgeois value systems.[14]

The "yuppie" phenomenon underscored how easily '60s individualism morphed into '80s materialism. The yuppie officially appeared in March 1983, discovered by *Chicago Tribune* columnist Bob Greene in an article about Jerry Rubin, the former yippie turned Wall Street networker. The following year, 1984, was dubbed "The Year of the Yuppie," in a *Newsweek* cover story. The yuppies helped define what personal expression meant in the '80s. Yuppies were obsessed with their material and professional advancement to the exclusion of other concerns. They openly flouted their philosophy through conspicuous consumption while taking a detached ironic view of civic affairs and social problems.

The media's discovery of yuppies in the 1980s seems quaint from today's vantage point. How odd that anyone would give a second thought to the materialism and naked self-interest of young urban professionals. But back in the early 1980s, when the yuppie first pulled up on the national scene in a gleaming BMW, the new individualism was in a transitional moment. It was still largely defined by the early baby boomers, whose notions of self-expression centered on artistic, sexual, and psychological exploration. The cultural firestorm around the yuppies was so intense because they hijacked the new individualism and took it in a sharply materialistic direction.

Criticism of the yuppies proved fleeting. A 1985 Roper poll showed that large numbers of Americans thought yuppies were "overly concerned with themselves," a view that reflected a wider public discomfort with how the new individualism was evolving.[15] But these sentiments had little traction in larger cultural debates. One reason was that no serious counterweight existed to the juggernaut of '80s materialism. Liberals were too busy worrying about Star Wars, the Contras, and Reagan Supreme Court nominee Robert Bork to attack the money culture. Also, somewhere along the line liberals had lost their ability to talk about values and their skills for moral storytelling. They spoke instead about "issues" and "constituencies." Among other things, this abdication allowed the right to successfully attack some parts of the new individualism while allowing other parts to flourish.

Neoconservatives and the Christian right teamed up to mount sweeping attacks in the 1980s on those aspects of individualism that clashed with family values, mounting a cultural war against sexual promiscuity, drug use, feminism, homosexuality, and artists like Robert Mapplethorpe. Conservatives had little to say—then or now—about the moral downsides of the money culture, such as greed, cheating, materialism, envy, and the ways in which careerism elbowed aside family and community. For all their invocations of God, it seemed that the right's moral missionaries had only read every other page of the Bible—ignoring the incessant warnings in both testaments about the evils of becoming obsessed with riches and growing callous toward the less fortunate.

From the early '80s on, the individualism spawned by the '60s evolved in a deeply lopsided way. Conservatives championed those individual freedoms associated with the free market, while deriding the hedonism associated with the counterculture. It became not all right in our society to express yourself by altering your consciousness with drugs or getting naked with strangers. But it was all right—admired, in fact—to express yourself with a Rolex, a Porsche, or a pedantic mastery of French wines.

For a few years, when the economy began to sour in the late 1980s, self-interested materialism experienced a setback. The stock market crash in November 1987 was heralded as the end of an era. "Drop Seen As Blow to Yuppies," read a *New York Times* headline. *USA Today* was less subtle in its obituary: "Sunset for Yuppies," it proclaimed. Hendrik Hertzberg published an essay in 1988 entitled "The Short Happy Life of the American Yuppie" in which he offered a scathing obituary: "We made the Yuppie into the effigy of selfishness and self-absorption, of the breakdown of social solidarity, of rampant careerism and obsessive ambition, of the unwholesome love of money, of the delusion that social problems have individual solutions, of the callousness and contempt toward 'losers,' of the empty ideology of winnerism and the uncritical worship of 'success.' Then we strung the little bastard up."[16]

Unfortunately the little bastard survived.

In hindsight, America's money obsession only hit a mild speed bump in the late 1980s on its way to a faster, more populist racetrack. Save for a freaky sliver of antimaterialist "downshifters," the money culture's conquest of America was complete by the end of the 1990s.

Opinion surveys confirm an explosion of material desires over the past two decades, along with a growing focus on financial success and the increasing linkage in people's minds between meeting these goals and achieving happiness. For example, in 1975, less than 20 percent of Americans surveyed identified a vacation home as being part of the "good life." By the early 1990s, that number had jumped to 35 percent. Less than 15 percent of people put a swimming pool in their dream scenario in 1975; in 1991, nearly a third of people did. A second car and a second color television became far more important during this period as well, as did travel abroad and having fashionable clothes.

Financial goals also began pushing aside other aspirations. The number of Americans who saw the good life as hinging on "a lot of money" and "a job that pays more than average" jumped substantially in the 1980s, rising from a minority of Americans who emphasized these two goals to a majority. As Americans expanded the list of material possessions that they saw as central to the good life, they de-emphasized the importance of other aspects of life. Even as more people admitted hankering after swimming pools and vacation homes, they also reported a declining focus on having a happy marriage or an interesting job.[17]

The shifting values and priorities of Americans over the past two decades have been especially evident among young people. Since 1966, the Cooperation Institutional Research Program has tracked the attitudes and demographic characteristics of college freshman. Over 700 campuses participate in the surveys every year. In all, more than nine million students have been surveyed. The trends among freshmen perfectly illustrate the rise of the money obsession and the sidelining of less materialistic goals.

In the first years of the survey, during the late 1960s, less than 50 percent of college freshman saw the goal of being well-off financially as either essential or very important, and just over half said the chief benefit of a college education is that it increased their earning potential. The shift toward materialism among young people began in the mid-1970s, as the idealism of the 1960s faded and harsh new economic times arrived. For example, the number of students who said they went to college to "make more money" climbed ten percentage points in just three years, beginning in 1976. By the mid-1980s, three quarters of college students stressed the goal of being well-off financially, a ratio that held through the

1990s. Today, this goal remains one of two or three of the highest-rated objectives of college freshman, along with the status-oriented goal of becoming "an authority in my field."[18]

By 1986, more than a quarter of college freshman reported that they planned to go into business, a career choice that trumped the closest rival by nearly three to one. This was also the year in which 40 percent of Yale's graduating class of 1,300 applied for a job at one investment bank alone, First Boston. Economics courses were oversubscribed at Yale and other top universities as students hurried to qualify themselves to join the gold rush on Wall Street. The promise of vast riches had an intoxicating appeal and was not illusory. "Never before have so many unskilled twenty-four-year-olds made so much money in so little time as did in this decade," wrote Yale graduate Michael Lewis, who joined Salomon Brothers. (This statement, of course, would be dated by the dotcom era.)[19]

As making money moved front and center, young people stopped caring about other things. In the late 1960s, believe it or not, the most important goal of college freshman was "developing a meaningful philosophy of life," cited by over 80 percent of entering students. The centrality of this goal waned steadily over the next twenty years, reaching an all-time low of 39 percent in 1987, at the height of the '80s boom. Interest in keeping up with politics and cleaning up the environment also markedly—and permanently—declined during the 1980s. The outlook among America's young was epitomized by Alex Keaton, the teenage star of the hit television show *Family Ties*. Alex was a briefcase-toting micro-yuppie, played brilliantly by Michael J. Fox. To the horror of his bohemian parents, he was all business and money, brandishing an inch-thick résumé in one episode. Alex never doubted that kids like him were the wave of the future.

He was right. While young people's interest in mainstream business careers dipped sharply after the crash of 1987, the money obsession morphed into a new kind of market populism by the late 1990s. Making money was way cool, and you didn't need a suit to do it. Teenagers swapped stock tips and traded over the Internet during lunch breaks. One fifteen-year-old in New Jersey amassed a nearly million-dollar fortune by masquerading as multiple people on Yahoo! and hyping low-priced stocks. The money was rolling in until SEC investigators arrived at his door.[20] Other teenagers built technology companies in their refinished basements, imagining themselves as the next Steve Jobs. If Alex Keaton came back to the future of the late 1990s, he would have staged an IPO from his bedroom.

For young people, though, the biggest social-health story of the 1990s was the onslaught of a virulent new strain of consumerism. The disease begins earlier and earlier with children these days, and it just gets worse. Parents complain endlessly about pressures from their kids to keep up with the Johnnies at the locker next door—with expensive video games, designer-label clothing, digital music players (to play pirated music), home computers, and cell phones. "Over the past 10 years, more people have come to think of themselves as having their identities shaped by their consumer goods," commented Alissa Quart, author of *Branded*, a book about consumerism among teenagers. "But teens and tweens are more vulnerable and more open to a warped relationship that the brands are selling to them. It's an emptied-out relationship where they pour themselves into a brand and see themselves through objects, rather than through people or ideas." Quart's book is filled with stories about sixth graders lusting after $500 Kate Spade bags and startling statistics, like how kids spend $600 billion a year of their parents' money and advertisers spend $12 billion a year to influence those buying decisions. She also shows the way in which movies and television programs about teenagers now emphasize an amazingly opulent lifestyle in places like Beverly Hills.[21]

A bit of competitive spending might be fine if it weren't so hard to keep up with those who set today's standards of material well-being. As Juliet Schor has documented in *The Overspent American,* nobody actually compares themselves anymore to the Joneses of yesteryear—that is, the next-door neighbor in a similar income group. We are now likely to compare ourselves with "reference groups" who make much more money than we do. If you're rich, you compare yourself to the superrich. You don't want a "McMansion," you want a real mansion—like the one you saw lovingly described in a rerun of *Lifestyles of the Rich and Famous.* If you're upper middle class, you compare yourself to the rich. If you're middle class or lower class, you might compare yourself to both the upper middle class and the rich.[22]

Stroll again into that Banana Republic and look around. Chances are that many of the young people buying the $42 polo shirts and the $78 jeans are plunking down plastic at what is anachronistically known as the "cash" register, and can't really afford these clothes. Unfortunately, in many urban areas the fashion taste of, say, a wealthy young advertising executive is akin to that of a low-paid editor at a food magazine. The editor shops at Banana Republic, in part, to fit into a social and cultural environment where she regularly compares herself to people who have five times her income.

With pressures like these, it's no surprise that many Americans believe that more money would make them happier. And yet, like a mirage, people's definition of enough money keeps flitting farther into the distance. Between 1987 and 1996, a period of only modest income gains for most households, the estimate among Americans as to how much annual income they needed to live in "reasonable comfort" increased by 30 percent, while the amount of money they felt they would need to fulfill all their dreams nearly doubled, from $50,000 to $90,000. According to one survey conducted in the early '90s, 85 percent of American households aspired to have a lifestyle associated with those in the top fifth of the income ladder. Only 15 percent said they would be satisfied ending up as middle class or just "living a comfortable life." A majority of Americans consistently report that they don't have enough income to live the life they want. Gaps between financial dreams and realities are hardly new in America. Some might even say that these gaps are what America is all about and that they are a good thing, keeping people on their toes and the economy humming. Yet there's a fine line between aspiration and envy, and between a healthy desire to get ahead and a relentless struggle to keep up. Judging by poll data, America crossed that line quite a while ago.[23]

Americans have pursued several solutions to rising financial pressures. First, they worked harder. Between 1979 and 2000, the amount of time spent at work by the average employee increased by 162 hours—an extra month a year. Between 1989 and 2000 alone, average annual hours increased by 95, more than two additional weeks on the job. Another solution has been to borrow money. There has been an explosion of credit card debt over the past twenty years. American households who have credit card debt have seen their balances increase by 66 percent in the past decade. Credit card debt among the lowest-income families more than quadrupled during this same period, with average balances rising from $594 to $2,440.[24]

Yet with even harder work and heavier borrowing, many people still can't keep up. Lower- and middle-income Americans are under the greatest money pressures—recall the ATM looters who worked full-time and yet had no savings. But persistent financial anxieties are also common among wealthier Americans. Because most people obsess about their relative well-being, comparing themselves to others nearby or on television, the line between financial "needs" and "wants" is easily blurred. Even those who have a lot of money can fret that they don't have nearly enough. This was particularly easy to do in the late 1990s, when the word "centamillionaire" was so often juxtaposed with "twentysomething" and "dotcom." *Newsweek* conjured the visceral anxiety felt by millions with a cover in early 2000 that shows a woman screaming: "Everybody's Getting Rich Except Me!"

This lament wasn't true, of course, but it seemed to be true at the height of the boom. Those who didn't have money felt worse about themselves during this era, while those who were wealthy felt more pressures to sustain and improve their position. Such pressures can be lethal to people's integrity. The fall of Jeffrey Silverman is a case in point.

Silverman was the prototypical child of privilege. His father was a multimillionaire corporate raider, and Silverman grew up in Forest Hills in the 1950s and '60s amid lavish surroundings. He was driven to private school every day in a limousine and groomed for success in business. Starting his career on home plate, he became at the age of twenty-one the youngest person with a seat on the New York Stock Exchange. He joined a company that made home-building products in 1981 and made $100 million when the company was sold in the late 1990s. He was initially less successful in love, divorcing twice, but by his fifties he was happily married again, to a beautiful heiress seventeen years his junior. The couple had twins.[25]

Silverman hoped to replicate his first success in business, and join the ranks of the truly rich, with another venture called BrandPartners. But things did not work out as planned. Entrepreneurship can be a good way to turn a large fortune into a small one, as the saying goes, and this is apparently what Silverman did. By 2000 he was hemorrhaging cash on BrandPartners as the company struggled. Silverman also had never fully paid taxes on the $100 million he earned from his earlier success and reportedly owed the

IRS over $30 million. In just a few years, his life had turned into a nightmare of impossible debts.

One big cause of Silverman's woes was his lavish spending habits. He owned a townhouse on Manhattan's Upper East Side, a second home in Palm Beach, and was building a third home in Bridgehampton, New York. He went to the best restaurants and friends said that he hadn't flown on a commercial jetliner in twenty-five years. All of this was confirmation of his public reputation as a very wealthy man who had made smart business choices and who was a player in New York's high society. And yet, by most accounts, Silverman was not a shallowly materialistic person. He was known as a man with a huge heart, who gave money to homeless people on the street and donated anonymously to various charities. He'd regularly tip the coat-check woman at his favorite restaurant, even when it was warm out. "Just because I don't have a coat, why should she suffer?" he explained to a friend. Silverman's life was filled with people who loved him intensely, and who relied on his warmth and counsel.

Money was not the driving force of Silverman's existence nor the thing that gave him meaning; it simply lubricated his life in every way. Appearances mattered deeply to Silverman, and when the financial pinch came, he was unwilling to give up these appearances. To deal with his financial problems, Silverman started stealing from BrandPartners. He granted himself improper compensation of at least $260,000 and also took reimbursements for personal expenses totaling another $130,000. The pressure and the cheating sent him into a depression, and his wife urged him to see a psychiatrist.

Silverman did not cover his tracks very well at the office. Soon people at the company were asking questions. Shareholders filed a lawsuit alleging misconduct by Silverman. By summer 2002, his life was unraveling fast. His debts were becoming unmanageable, and his lies and stealing were on the verge of being exposed at a moment when corporate villains were being pilloried in the press as rarely before.

Many high-profile Americans in recent years have endured terrible humiliation and bounced right back. This is not hard to do in a culture that loves good confessions and indulges second acts. Silverman could have pulled through. Prison time would have been unlikely and, with the right moves, he could have held on to many of his assets. In the worst case, he and his wife could have lived well off of her family's wealth.

Silverman did not prove so resilient. One afternoon in September 2002, he left a voice message for his wife saying "I love you very much, I love the children very much," but explaining that he couldn't go on anymore. Later, the Greenwich police department received a phone call from someone reporting a distraught man down by the Long Island Sound. When the police arrived, they found Silverman dead from a gunshot wound to his chest.

ON THEIR OWN, extreme individualism and the money obsession might be expected to bring out the worst in Americans. Yet these traits have been amplified by another self-interested outlook associated with laissez-faire ideology, social Darwinism. Americans, never sympathetic to life's losers, seemed to grow even more callous toward the weak during the '80s and '90s. We also turned more worshipful of the strong—namely, the rich and famous. Why are these attitudes relevant to cheating? Because it is a small step from believing that big shots intrinsically deserve their power to indulging the abuse of that power.

In fairness to Charles Darwin, it should be said that the great naturalist never applied his evolutionary theories to modern human relations. We can thank the British philosopher Herbert Spencer for that clever twist. Spencer, a leading conservative thinker of the nineteenth century, actually coined the term "survival of the fittest" before Darwin published *On the Origins of Species* in 1859. Spencer used ideas about individual freedom and society's "natural" competitiveness to argue against assistance to the poor or state interference in the economy. The publication of Darwin's scientific writings provided Spencer and his followers with proof that their harsh views of commerce did indeed reflect a morality derived from the inarguable truth of nature. Some people were just naturally suited to rule and prosper, others to be ruled and fall by the wayside.

Social Darwinism proved popular among the industrial elites of the late nineteenth century. In America, it penetrated deeply into politics and culture and was used to rationalize the inequities of the Gilded Age. It was not until the early twentieth century that pragmatist thinkers fully debunked social Darwinism as a legitimate way to talk about economic life. But social Darwinism continued to find adherents. It was used by Jim Crow racists to justify white hegemony in the South and found new supporters in the eugenics movement. Imperially inclined American elites also cited a natural order of nations in rationalizing U.S. colonial holdings in the Philippines and Caribbean.[26]

Social Darwinism held less sway in American life during the national emergencies of the Great Depression, World War II, and the Cold War. There was more of a sense in those decades that Americans were "in it together," and that everyone had an important role to play. The economic Great Compression of the 1950s and 1960s brought equity norms into business and helped the U.S. move toward manifesting its historic destiny as a classless society. Also, the Civil Rights movement and then the women's movement challenged those notions of innate superiority that endured in American life.

Still, certain ideas associated with social Darwinism have a perennial foothold in U.S. society—you might even say these ideas are part of our national DNA. Contemporary social Darwinism is not like that of our great-grandparents' day. Conservatives and libertarians don't openly invoke natural selection in explaining their opposition to food stamps or why it's okay for Wal-Mart to take over the nation. And, with the exception of Charles Murray and a few others who argue that some races are naturally smarter than others, it's rare to hear justifications for inequality that hinge on biological determinism.

Today's social Darwinism is heavy on the social and light on the Darwinism. It starts with the premise that America remains a true land of opportunity, an equal playing field where everyone has a shot at success. Those who succeed in a big way have proved that they are smart or hardworking or talented; they rightly deserve all the rewards given to them. Polls show that large majorities of Americans believe that anyone who works hard can succeed, and even higher percentages of Americans say they admire people who get rich by their own efforts. Those who fall behind, meanwhile, are often blamed for their misery. In a typical recent survey finding, three quarters of Americans agreed that "most people who don't get ahead should not blame the system, they have only themselves to blame." In 2001, nearly half of Americans agreed with the statement that if a person is poor, their own "lack of effort" is to blame.[27]

In other words, Americans tend to make moral judgments about people based upon their level of economic success. Everybody loves a winner, the saying goes, and nowhere is that more true than in America. Winners are seen as virtuous, as people to admire and emulate. Losers get the opposite treatment—for their own good, mind you. As Marvin Olasky, an adviser to President George W. Bush, has said: "An emphasis on freedom should also include a willingness to step away for a time and let those who have dug their own hole 'suffer the consequences of their misconduct.'"[28] The prevalence of a sink-or-swim mentality in the United States is unique among Western democracies, as is the belief that individuals have so much control over their destiny. Elsewhere people are more apt to believe that success or failure is determined by circumstances beyond individual control. Scholars attribute the difference in outlook to the "exceptionalism" of America and, especially, to the American Dream ethos that dominates U.S. culture—an ethos at once intensely optimistic and brutally unforgiving.[29]

All of this has troubling implications for our society's ethics. Americans reflexively cut slack for those who are successful. We may admire winners whatever their sins. As sociologist Robert Merton wrote fifty years ago, the "sacrosanct goal" of wealth "virtually consecrates the means"—any means.[30] F. Scott Fitzgerald's Jay Gatsby is an iconic figure in this regard—he was irresistibly appealing despite the sordid origins of his fortune. Real-life America has been filled with similar characters, ranging from Joseph P. Kennedy, who made much of his fortune illegally, to Michael Milken, who easily rehabilitated his image following his conviction for insider trading in the late 1980s. Even very nasty people who prevail in ugly mudslinging or backstabbing contests can win grudging respect—as Richard Hatch did when he triumphed on the reality television show *Survivor.*

Survivor was supposed to simulate a situation in which a group of sixteen people must battle the elements and eke out an existence on a desert island off Borneo. But the show was intentionally less *Gilligan's Island* and more *Lord of the Flies.* The true challenge was to remain the last person on the island as a series of votes were held among contestants and people were voted off the island one after the other. The winner received $1 million.

When all the cheating and conniving on the island were finally over, the winner of the jackpot was Hatch, a corporate trainer. Modern management techniques, it turns out, are pretty helpful if you're trying to make it on a cutthroat island.

Was Hatch seen as suspect for triumphing in a snake-pit environment that brought out his and everyone else's worst side? Not a chance. He was feted by the television talk shows and offered various moneymaking opportunities. Nobody cared how Hatch won. They only cared *that* he had won.

At the same time, our lack of sympathy for life's losers makes us less prone to outrage when ordinary people are exploited by powerful cheaters. We are as likely to blame the victim as the villain. For example, in the aftermath of corporate scandals, a number of commentators pooh-poohed the losses of those who invested in Enron, WorldCom, and other corrupt companies, saying they should have known better. "Investors who lost their life savings did so because they made a stupid investment decision," commented one college student after Enron went belly up. "They failed to follow the common sense rules of asset diversification."[31] This view was widely echoed by media pundits, even as it became clear that Enron employees had been locked into 401(k)s loaded with company shares—while top executives dumped their own stock.

Corporate scandals offer powerful evidence of the corrupting impact of the value shifts of recent decades. Like Scott Sullivan, many of the executives caught up in these scandals are not much different in background from other Americans. That they ended up as big-time criminals isn't due just to lax regulators or huge financial temptations. Corruption in the executive suite is also explained by the moral climate of corporate America, a place where the troubling value shifts in our society have played out with notable intensity.

Enron's corporate culture in particular will endure as an archetype of bad values in high places. The company is a vivid example of what can happen when you stir together the leading moral toxins of the '90s—extreme individualism, money obsession, and social Darwinism.

At the height of its reign as America's "most innovative company," Enron's corporate headquarters in Houston was filled with employees that, on the surface at least, looked pretty much alike. In the futuristic fifty-story glass tower where Enronians worked, the "Death Star" as it was called, an informal dress code held sway. Men generally wore khakis and, invariably, a blue shirt. Most had buzz cuts and goatees were common. It was a young workforce, recruited from America's best business schools, and a remarkably fit and attractive crowd. The employees of the Death Star could have passed for a hand-selected master race preparing to journey into space to colonize another galaxy. Indeed, the men who ruled the executive suites of the fiftieth floor—Jeffrey Skilling (aka "Darth Vader") and Kenneth Lay (aka "the Emperor"), tended to imagine themselves as colonizing and commodifying one area of economic life after another with their trading models.

The most profitable part of Enron's operations sought to predict future price fluctuations for a wide range of items and then placed bets based on those predictions. "Enron was, in reality, a derivatives-trading firm, not an energy firm," explains Frank Partnoy in *Infectious Greed*. This trading was "highly profitable, so profitable, in fact, that Enron almost certainly would have survived if key parties had understood the details of its business."[32]

Enron's leaders believed they were reinventing the corporation by creating a company that wouldn't actually make anything—except profits as it took advantage of the ebb and flow of fast-changing markets in which everything under the sun was bought and sold. Free markets were an article of faith at Enron. The freer the market competition and the less the regulation, the more that Enron could enter the picture and make money. For example, California's deregulated energy market was the ideal hunting ground for

Enron traders. It also illustrated the ways in which free markets could turn corrupt as Enron used its powerful position in the energy market to drive up prices. Millions of California consumers paid higher energy bills as Enron raked in higher profits.

Enron invested heavily in politics in its drive to see wider deregulation and freer markets. Enron was one of the largest contributors to George W. Bush's 2000 election, but it also gave money to many members of Congress on both sides of the aisle.

As Enron used its political clout to push for a new era of free markets, Jeffrey Skilling sought to build a workforce that had the creativity and flexibility to thrive in an environment of relentless competition. He envisioned Enron's workforce as something of a superior breed, carefully selected through a rigorous screening process. While Enron's fit and youthful employees may have looked like a generic new economy army marching in perfect step, in actuality Skilling filled Enron with quirky individualists and cunning entrepreneurs who were not encouraged to cooperate with one another. Before its fall, Enron had one of the most cutthroat cultures in all of corporate America.

A 1975 graduate of Harvard Business School, Skilling was recruited to Enron by Kenneth Lay in 1990. He came over from McKinsey & Company, a consulting firm known for its brilliant partners and a nearly religious fervor for bottom-line methodologies, which it preached on a global scale. Skilling was the driving force behind Enron's transformation in the 1990s. He became chief operating officer in 1997 and CEO in 2001. Over the years, Skilling lost his pudginess, along with his white-collar pallor. He got laser eye surgery, left his suits in the closet, and took up extreme sports. He turned physically fit and notoriously focused, not unlike a bullet in his persona. "Skilling was known to himself and others as the smartest human being ever to walk the face of the earth," recalled Brian Cruver, a former senior manager at Enron.

"He never lost, and he never failed. He was arrogant and ultra-competitive."[33]

Skilling actively sought out highly intelligent and iconoclastic people to join Enron. "I always said our weirdest people are our best people," Skilling once explained. "Bring on the weird people." At the same time that he created a new breed of employee, Skilling mounted a merciless campaign to exterminate Enron's old guard, the employees who had joined Enron when it was still a traditional energy company. He abolished seniority preferences in favor of a system that offered huge cash bonuses and stock options to top performers. He also tried to get rid of any structures that might stifle individual creativity or initiative. He was against staff meetings and down on status reports and opposed to dress codes and not keen on strategic planning.[34]

Skilling sought to shape Enron into something akin to a pure state of nature, where individuals could advance themselves—and the company stock—unencumbered by the typical restraints found in corporate civilization. Enron was a place where a fresh-faced MBA could make trades worth millions of dollars without higher approval.

It was also a place where people learned to watch their backs.

Vicious competition among employees was a natural outgrowth of the Skilling approach. Traders were sometimes afraid to go to the bathroom, for fear that their neighbor might steal information off their computer screen. When they went home at night, traders locked their desks so that nobody would rifle through the contents. Skilling encouraged the dog-eat-dog culture, and it reflected his own survival-of-the-fittest mentality. In his rare downtime, Skilling liked to go on risk-filled adventures, such as desert motorcycling trips in Mexico or scavenger hunts in the Australian outback or safaris in Africa. He was known to say that his ideal excursion was so dangerous that someone could die—although not himself, presumably.

The equivalent of natural selection at Enron was the performance-review system that Skilling implemented within the company. Better known as "rank and yank" or "forced ranking," the system compelled Enron managers to rate all their employees every six months from one to five, and then fire 15 percent of those at the bottom every year, even if these "fives" were actually not bad employees. The review system also allowed any Enron employee to give input on any other employee.

Rank and yank became common in the ruthless corporate world of the 1990s, and remains widely used today. Jack Welch, the former leader of General Electric, was the most vocal apostle of rank and yank. Every year, GE ranked its entire labor force and then fired the bottom 10 percent. "GE leaders must not only understand the necessity to encourage, inspire and reward that top 20 percent," Welch said in 2000. "They must develop the determination to change out, always humanely, that bottom 10 percent, and do it every year." Welch—who proved his own fitness by seducing the sexy young editor of the *Harvard Business Review* and finagling a bloated retirement package worth around $150 million—never wavered from this belief. One day while shopping on Fifth Avenue, he was pulled aside by an anxious store manager. "I've got 20 salespeople," the manager said. "Do I have to fire two of them?" Welch told him he did indeed.[35]

The ruthlessness of rank and yank conjures up the sci-fi arenas of *Logan's Run* or *Brave New World*, although even in Aldous Huxley's hierarchical dystopia those at the bottom—the Epsilons—were allowed to stick around as servants of the Alphas and Betas who ran the show. The inferiors at places like GE and Enron were shown no such compassion. At Enron, Welch's emphasis on "humanely" excising the bottom tier of employee got short shrift, and Enron was known to have one of the harshest rank-and-yank systems in corporate America. Those employees ranked as ones often got seven-figure bonuses. Employees ranked as fives were "redeployed" to a special area in the Death Star, where they sat among themselves at workstations and had a few weeks to look for another job. Theoretically they were eligible for other jobs at Enron, but few managers wanted to hire a five. "The whole thing just made me sick," Brian Cruver said later. "The whole 'rank and yank' process seemed more like 'displace and disgrace.'"[36] Maybe, in fact, humiliation was part of the point.

While Enron's rank-and-yank system was supposed to create a brutal but effective meritocracy, things didn't work out that way. The system was frequently used for vendettas and people's ratings were as likely to reflect their skills at office politics as anything else. Robert Bruner, a business professor who studied Enron extensively, commented that "rank-and-yank turned into a more political and crony-based system." Managers were known to lie and alter personnel records to get rid of certain employees. Andrew Fastow, Enron's chief financial officer, was notorious for using performance-review sessions to retaliate against those employees who had dared to cross him. Other managers at Enron also used the forced-ranking system to silence dissent. The distrust and rivalries engendered by the rank-and-yank system, along with the company's focus on individual accomplishment, created an environment at Enron antithetical to team building. It was a place that revolved around a kind of daily emotional violence. Abuses were inevitable, especially given how much money was at stake.[37]

And money was what life at Enron was all about. Making money for the company, and yourself, was the overriding criteria by which everyone at Enron was judged. Enron was filled with managers who had left other companies that had loftier, more complex notions of employee success in order to seek their fortune in Houston. There was no cap on the size of cash bonuses that could be paid to Enron employees. If you were good, you could get rich beyond your wildest dreams.

In 1998, when Enron's stock passed $50, every employee found a crisp $100 bill waiting on their desk the next morning. In Enron's London headquarters, a banana-yellow BMW Z3—the kind of car James Bond drove in *Goldeneye*—sat in the lobby for months, as a prize for an employee who referred new recruits to Enron and then won a raffle. Tacky? Perhaps at ExxonMobile or Microsoft. Not at Enron. Maybe it was the culture of one-upsmanship, maybe it was the unlimited bonuses, or maybe it was because Enron was in Texas, a notorious incubator of nouveau riche excesses, but conspicuous consumption reigned at Enron with an intensity uncommon in *Fortune* 500 companies.

Top executives set the tone. Ken Lay, who openly admitted his aspiration to be "world-class rich," lived with his second wife (and former secretary) in a $7 million penthouse in Houston's exclusive River Oaks neighborhood. He owned a dozen other properties beyond the penthouse, including four houses in Aspen. His real estate buying spree barely put a dent in the personal fortune he had amassed by selling $144 million in Enron shares when the stock was still worth something. Lou Pai, who had met his second wife at the Men's Club in Houston, exemplified what retirement might look like if you played your cards right. Pai cashed out of Enron with $350 million that he used to buy a 77,000-acre ranch in Colorado, along with a horse-breeding ranch outside of Houston. Pai wasn't the horse freak; his young stripper wife, Melanie, was.

Enron executive Ken Rich neither collected horses nor picked up strippers, but he did have a weakness for cars. Once he appeared at an Enron outing in the Texas hill country with a whole trailer truck of his Ferraris. Skilling, who'd also traded in his first wife to hook up with a corporate secretary at Enron, Rebecca

Carter (nicknamed "Va-Voom"), followed Ken Lay to the River Oaks 'hood, where he bought a $4 million Mediterranean-style mansion. The price tag was small change compared to the $76 million Skilling made by unloading shares of Enron stock before the company went bust, and there was plenty of room for the two of them in the sprawling home. Carter fit easily into the rich neighborhood, since she was rich herself; Skilling had raised her salary to $600,000 a year.[38]

The wealth of top executives no longer stands today as incentive for working hard and skewing earnings reports at Enron. However, while Enron is now bankrupt, its former executives aren't. Neither Lay nor Skilling has yet to be fined a dime for what happened at Enron, much less indicted. Skilling still lives in his mansion, with his huge fortune largely intact.

And Lou Pai still owns his ranch in Colorado. Among other things, the huge spread includes a 14,000-foot mountain peak. Many in the region have dubbed it "Mount Pai."

CHAPTER FIVE

Temptation Nation

T*HE BOOK OF VIRTUES* IS THE KIND OF BESTSELLER PHENOMENON that comes along only a few times a decade. Published in 1993, it is a mammoth collection of "moral stories" edited by William Bennett, who served as Secretary of Education under Reagan and drug czar under George Bush. For eighty-eight weeks *The Book of Virtues* held a spot on the *New York Times* bestseller list and helped establish Bennett as the premier moral commentator of the 1990s. This book and others, along with frequent lectures, have made him a rich man—rich enough to gamble away some $8 million at casinos and still have plenty of money. Bennett says his gambling days are over, not that this is any of our business. Now, when he needs downtime, Bennett often heads south from his home in Washington, D.C., to a beach house on the Outer Banks of North Carolina—a place he sometimes calls "the house that virtue built."

Like Bennett's other books—such as *The Moral Compass* and *The Devaluing of America*—the themes of *The Book of Virtues* track with conservative jeremiads of the past two decades: that America has lost its moral way because of a decline in traditional values like self-discipline, religious faith, and devotion to marriage. Bennett and other conservatives lay blame for this moral crisis on bad liberal ideas and a paternalistic government that has usurped the traditional roles of family and civil society. (The religious right adds proliferating casinos to this list of corrupting influences; Bennett, obviously, has a different opinion.)

Bennett—himself a former liberal who once had a blind date with Janis Joplin—is rare among conservatives in that he has occasionally targeted the free market for criticism. "There's obviously a tension between the market and virtue," Bennett once said. "The market is all about creating desire and gratifying it. Virtue is mostly about postponing gratification." These criticisms, however, have been muted, and Bennett hasn't had much to say about corporate scandals or the other areas of cheating in American life related to money and career.[1]

There is, in fact, little in the broad conservative critique of America's moral decline that can explain the rise of cheating across U.S. society in the past quarter of a century. Even as conservatives warned us through the 1990s about the pathologies of single mothers on welfare and the dissipation of young people with drug habits picked up from their pot-smoking baby-boomer parents, they ignored the negative effects of intensifying competition for money and status across all sectors of society. One reason that the corporate scandals took America by surprise is that conservative diatribes about the "cultural war" directed attention away from the morally corrosive potential of extreme capitalism. Thus distracted, Americans weren't expecting the many ugly excesses that seem, in retrospect, to have been inevitable.

For example, in his 1996 bestseller, *Slouching Toward Gomorrah*, Robert Bork bemoaned the decline of America culture within a framework better suited to the '70s than to the '90s. Bork charged that "the enemy within is modern liberalism," that the dangerous left-wing orthodoxies of "radical egalitarianism" and "radical individualism" were bringing out the worst in Americans. The sins that concerned Bork included crime and drugs and illegitimate children, as well as "feminism, homosexuality, environmentalism, animal rights—the list could be extended almost indefinitely." Quite apart from the corrosion of family and community, Bork worried that liberal orthodoxies would destroy America's economic vitality. He wrote that "it seems highly unlikely that a vigorous economy can be sustained in an enfeebled, hedonistic culture, particularly when that culture distorts incentives by increasingly rejecting personal achievement as the criterion for the distribution of rewards."[2]

What country was Bork living in? While Bork's book made some astute observations about the pull of radical individualism in America, he got just about everything else wrong. Egalitarianism didn't thrive in the 1990s; it was pushed further to the sidelines of U.S. society as harsh norms of economic competition took center stage. And the most potent fuel for bad behavior in the '90s was not the temptations of sexual pleasure or altered conscious, much less indulgent sympathies for baby seals or redwoods; it was the lure of wealth, status, and luxury. Like most conservatives, Bork was so enthusiastic about the free market that he didn't grasp the ways in which an escalating money culture could corrupt even the most upstanding citizens.

It's now more obvious than ever that being moral in narrow conservative terms is no protection from sinning in other ways. Take someone like Bernard Ebbers. As head of WorldCom, Ebbers was known as one of the most religious CEOs in the high-tech sector. He invoked God regularly in speeches and press interviews, and started each board meeting with a prayer. He was a deacon at his Baptist church, where he also led a weekly bible study class. According to those who took the class, he was remarkably fluent in scripture. If someone missed the class, Ebbers would be on the phone to see if they were okay. And yet, Ebbers presided over the largest fraud in U.S. history, a fraud that wrought massive financial pain on present and future retirees across America. After the revelations of this crime, a tearful Ebbers told his congregation: "More than anything else, I hope this doesn't jeopardize my witness for Jesus Christ."[3] Ebbers is now under indictment in Oklahoma, but has yet to face federal charges.

Or look at John Rigas, who headed Adelphia Communications, one of the largest cable television companies in the United States. The son of Greek immigrants, Rigas was a regular churchgoer guided by social conservatism. He raised his four children in a small town in upstate New York with a strict set of traditional values. His sons went to work with their father after graduating from the nation's best colleges and they, too, became pillars of the upstate community where Adelphia was headquartered. The sons, like the father, were social conservatives. No porn channels were allowed on Adelphia's cable system.

A very different morality guided the Rigases in business. By the time investigators caught on, the Rigases had appropriated hundreds of millions of shareholder funds for their personal use through various shady loans and frauds. Prosecutors accused father and sons of "systematically looting" Adelphia; the Rigases are currently fighting the indictment.[4]

Philip Anschutz is yet another business leader who publicly embraced religion and "family values"—while indulging in greed and financial chicanery at the office. A billionaire who is the largest owner of movie screens in America, Anschutz is a religious man who has crusaded against homosexual rights and the medical use of marijuana. He has bankrolled a variety of Christian conservatives and invested in prayer radio. Yet as the founder and chairman of Qwest Communications, a telecommunications firm, Anschutz ranks among the most corrupt insiders of the late 1990s. He sold nearly $2 billion of Qwest stock as it plunged in value from $63 a share to $3. As these sales took place, many in a secretive fashion, Qwest was encouraging its employees to hold on to their own stock and to build their retirement plans around 401(k)s heavy with Qwest shares. Anschutz was later investigated by Eliot Spitzer's office and eventually agreed to give up $4.4 million in illegal gains from his shady business dealings without admitting any wrongdoing.[5]

As LEAD ADVOCATES of downsizing and defanging government, conservative politicians and intellectuals helped create the kind of permissive environment where corporate scandals occurred. Even as they claimed to be tackling America's "moral crisis," the Borks and Bennetts ignored the temptations that are naturally part of capitalism and pushed policy ideas that raised these temptations to a new level. Regulatory checks shriveled at the same time that a booming economy increased the rewards for lying and cheating. The SEC found that it lacked the resources—and legal authority—to police increasingly complex financial markets, while an underfunded IRS was outgunned by ever-more-sophisticated tax cheats. During the deregulation craze the government scaled back rules governing utilities, banking, telecommunications, airlines, trucking, and other industries. A zeal for privatization led to more government functions being turned over to private contractors—often with few safeguards to prevent abuse. The flood of money into politics, deeply corrupting both parties, partly explains why government enforcers were sidelined during the boom. Free-market ideology explains even more.

Cheating thrives in an era of big loopholes and drugged watchdogs. If you run a business that generates waste, you might be more tempted to spoil the environment because you know that strapped EPA enforcers aren't going to bother with you—or because a friendly administration has changed the law in your favor. If you're selling power during an energy crunch, you might be tempted to manipulate the energy market and artificially drive up electricity prices in order to make a killing. Why? Because you can—because deregulated energy markets and unsophisticated enforcers allow you to get away with profiteering. If you're a landlord in New York City, with its perpetual housing crisis, you'll be tempted to cut all sorts of corners because you know that these days the city has the resources to send out investigators only if there's a flood or a collapse, leaving much illegal behavior unpunished.

If you're Wal-Mart, America's biggest company, you'll be tempted to continue the pervasive practice of forcing workers in your stores to put in overtime without paying them for it or using illegal immigrants. While your friends in Washington have not yet closed down the Department of Labor, they've at least kept it on a starvation diet for the past twenty years, leaving its investigators outgunned. If you are fined, the damage is sure to be less than the money you've made by flouting the most elementary of American labor laws.

And if you're an accountant like Michael Conway, you may be tempted to let your clients cheat their shareholders by telling multibillion-dollar lies.

Conway lives in Westport, Connecticut, an exclusive suburb known for celebrities such as Martha Stewart and Robert Redford. Outside of the accounting world, nobody's ever heard of Conway. Inside that dry and arcane world, everyone has. Conway is a longtime senior partner at KPMG, one of the nation's largest accounting firms. He has served on a variety of boards and committees within accounting associations, as well as at KPMG, where he is a member of the firm's partnership board and chairman of its audit and finance committee. As a respected leader in the field, Conway has helped shape accounting policies and practices, both in his own firm and at the national level.

Lately, though, Conway has been getting the kind of attention that no accountant wants. He's embroiled in a major legal mess stemming from SEC charges and could face prison time on multiple felony counts. His days are often spent cloistered with lawyers.

The story of how Michael Conway ended up in the crosshairs of SEC investigators has become a familiar one in recent years. Conway and three other senior members of KPMG are charged with playing dumb while the Xerox Corporation wildly overstated its earnings, thus misleading investors about the company's profitability. Xerox did this for the same reason that numerous other companies misstated their earnings during the 1990s and into the twenty-first century: to meet its "performance expectations" on Wall Street and thus boost the price of its stock.

Xerox's top executives, like many leaders of big companies, had an urgent financial interest in the stock price thanks to the widespread practice of compensating executives with large numbers of stock options. In theory, stock compensation is supposed to help align the interests of company leaders with those institutions and individuals who own stock in the company. At Xerox and elsewhere, it had the opposite effect. It gave corporate leaders reason to cheat and mislead investors in the quest to rack up huge personal fortunes. And, in case after case, the accountants serving as watchdogs had their own financial motives for looking the other way.

Xerox has one of the best-known brand names in corporate America. The company shot to the top ranks of big business during the 1960s, after it introduced the 914 office copier. Peter McColough, the CEO who charted Xerox's rise, believed that large corporations had a responsibility to do good even as they did well, and distinguished himself as among the most socially responsible business leaders of the 1970s.

Xerox's powerful brand identity helped it through tough times in the '80s, and the company had rebounded by the late '90s. One might think that the stewards of the hallowed Xerox name would protect the company's reputation with their lives. Instead, according to the SEC, company leaders focused obsessively on protecting Xerox's stock price and their own fat pay packages. When the stock price was in jeopardy, they engineered a far-reaching fraud to misreport Xerox's earnings. The fraud was complex, as these schemes tend to be, and involved misreporting revenues generated by Xerox's leases of copying machines around the world. Ultimately, according to the SEC, "the company defrauded its shareholders and the investing public by overstating its true equipment revenues by about $3 billion and its true earnings by approximately $1.5 billion" between 1997 and 2000.[6]

This large-scale fraud, investigators say, would not have been possible without the complicity of Michael Conway and his colleagues at KPMG.

Auditors are supposed to prevent fraud from occurring, not facilitate it. They are among the most important sentinels within modern capitalism, charged with keeping business numbers honest. Without reliable financial data, investors both large and small cannot have confidence that they are putting their money in a safe or profitable place. Quite apart from the havoc that bogus numbers can wreak on the economic well-being of individual stockholders, at stake is the broader functioning of the economy. Without investor confidence, money flees financial markets and growth slows.

For all these reasons, accountants have solemn obligations to be honest. Their professional code of conduct—as well as the terms of their government licenses and of federal securities law—obliges them to follow a set of established rules known as the Generally Accepted Accounting Principles (GAAP).

Michael Conway and KPMG ignored the rules when they signed off on Xerox's cooked books, according to the SEC. "The

KPMG defendants were not the watch dogs on behalf of shareholders and the public that securities laws and the rules of the auditing profession require them to be," the SEC charged. The specifics of the SEC charges revolved around Xerox's use of "self-serving, untested assumptions, and improper accounting methods" to boost earnings, and the willingness of KPMG to endorse these methods, even though they violated basic standards of accounting.[7]

Michael Conway was not part of the scheme when it began in 1997, but he signed off on a 2000 audit of Xerox that allegedly perpetuated the bogus accounting. Conway was brought to the Xerox account in 2000 after another KPMG accountant, Ronald Safran, was taken off the job because Xerox's top executives complained about him to KPMG chairman Stephen Butler. Safran, who lives in the leafy suburb of Darien, Connecticut, is a hard-working CPA who has been with KPMG since graduating college in 1976. His sin was that he had questioned Xerox's management about its fraudulent accounting practices. It was the second time in six years that KPMG's lead auditor for Xerox had been sacked. To keep Xerox happy and to keep a lid on questions about the company's cooked books, Butler turned to a trusted senior lieutenant in the form of Michael Conway.

Conway had been warned by Safran about Xerox's dishonest numbers. Safran told him that Xerox was scurrying at the end of reporting quarters to "bridge the gap" between the earnings it promised and the earnings it achieved. Safran said that something had to be done to raise a red flag. Nothing ever came of these warnings.

Conway did more than just sign off on false numbers for 2000, according to the SEC. He also participated in the cover-up that followed. In early 2001, when KPMG learned that the SEC investigation was under way, it advised Xerox to restate its earnings.

Xerox hired outside consultants who came up with a new set of numbers aimed at appeasing the SEC—but the numbers were still fraudulent. KPMG then endorsed these new numbers, even though they had no reason to believe that they were legitimate. It wasn't until KPMG was replaced later in 2001, and Xerox went along with a truly legitimate audit, that an accurate restatement of earnings occurred. The new numbers made headlines; at the time they were the largest restatement of earnings in U.S. history. And this latest episode of brazen fraud had not occurred at some company down South filled with high-tech cowboys—it involved a blue-chip pillar of America's eastern business establishment.

The SEC's sweeping charges against KPMG for its role in the massive Xerox fraud were filed less than a year after the restatement of earnings. The complaint was hardly the first time that KPMG had caught the attention of the SEC. The firm has been embroiled in countless scandals over the years. For example, it paid a $9 million settlement in 2001 for its role in facilitating a $600 million Medicare fraud by the health-care giant Columbia/HCA. The Xerox complaint was also not the first time that Michael Conway had been slammed by the agency. Just two years earlier, the SEC had reprimanded Conway and KPMG for compromising their auditor independence in unsavory dealings with a firm called Porta Systems Corp. Conway, the SEC concluded, was "alarmingly careless in not learning what he needed to know and in not appreciating the significance of what he knew." The Porta case never resulted in federal charges.[8]

Not surprisingly, KPMG tells a different story of its relationship with Xerox and about Michael Conway's professional integrity. It praised Safran for standing up to Xerox and said he had forced the company to make a number of adjustments to its financial statements. Safran was taken off the account not for being too tough on a company committing fraud but rather because of

"Xerox's professed inability to communicate effectively" with him. (Xerox: "Back off, pal." Safran: "Que? No comprendo.")

And, says KPMG, Michael Conway wasn't brought in as a reliable cover-up man; he came in and tried to make Xerox clean up its act, forcing them to get real with their numbers in the "face of massive client resistance." Conway was a hero in the KPMG story, and KPMG itself said it "did the right thing. We stood up to the client and asked the tough questions.... It is astonishing to us that the SEC would choose to bring this action."[9]

In June 2003, six former Xerox executives settled with the SEC, agreeing to pay $22 million but—you guessed it—admitting to no wrongdoing. The SEC and KPMG remain in litigation. Yet assuming that the SEC's charges are correct—and that this badly overstretched agency wouldn't pursue a case that it wasn't certain of winning—one naturally asks: What was KPMG thinking? Why would it go along with massive cheating? Why would Conway, a distinguished leader in his field, sign off on numbers that were outrageously bogus?

None of this is much of a mystery. Xerox was a top client of KPMG, which had been the company's auditor for forty years. During the four years that KPMG turned a blind eye to Xerox's fraud, it earned $26 million in auditing fees from the company. This stream of revenue itself might have been worth lying to preserve. But KPMG billed Xerox more than twice as much in the same period, $56 million, for a wide range of consulting services. Nobody at KPMG was willing to stand up to Xerox and place this revenue stream in jeopardy. Hefty year-end bonuses were at stake for partners like Conway, who had long been involved in nurturing the relationship with Xerox. When Xerox turned bad, everyone had every reason to look the other way. See no evil, hear no evil—make a lot of money.

Life at a big accounting firm can be extremely stressful. While partners worry about their bonuses, accountants down the ladder worry about their jobs. The pressures to play ball in unseemly situations can be immense. Erik Hille, a former manager at KPMG, describes the situation this way: "The question is how far do you want to push each rule? How many loopholes do you want to find? Nobody says this firm has an aggressive or conservative approach to their audit on the front page of their disclosure. The new hire has no power to say 'no.' Their job is on the line to get a 'yes' answer.... The new hires fear very strongly for their job."

The dance between auditors and clients is often very delicate, especially when corporate managers are under intense pressure to put out positive earnings numbers. William Ezzell, a partner at Deloitte & Touche, and a leading figure in the accounting profession, describes the interaction around a company's skewed numbers this way: "Sometimes they just do it without consulting the accountant, and if the accountant raises a question you're at a pressure point—a point where they've convinced themselves they're right. From my personal experience, the dialogue that takes place is 'You need to help me get it right.' But what is meant is 'Help me figure out a way around the rule.' Or, 'How can I do what I want to do? If this isn't right, give me another answer to get to the result I want.'... They will often say, 'You've got to be more aggressive.'"

Like corporate law firms, accounting companies have been transformed in the past few decades by greed and a harsh bottom line. Life has become far more insecure at august accounting firms that no longer offer most young associates a good shot at making partner. "There are few partners at the top and the promotion to become partner is very stiff," comments Erik Hille. Yet partners, too, have begun to feel much more heat as they constantly try to prove themselves and justify large bonuses. Partners are "under pressure to keep their revenue up and to keep their clients happy," says Hille. "A large percentage of the partners' income is based on the revenue they generate." The more money at stake with a client, the more auditors might be willing to go along with bogus numbers. And in the 1990s, as accounting firms increasingly ventured outside the auditing business, consulting for the same companies they audited, the financial stakes in client relationships rose dramatically.[10]

KPMG's ties with Xerox exemplified these profound conflicts of interest. This recipe for cheating existed despite constant warnings from government regulators. But with a Democratic president declaring in the 1990s that "the era of big government is over," with Washington awash in special interest money, and with Congress controlled by conservative ideologues, government regulators found themselves largely impotent as the most basic safeguards of modern capitalism were removed.

ARTHUR LEAVITT IS NOW famous as the man who cried wolf about big bad accountants throughout the 1990s. Each time, Congress was paid handsomely not to listen.

Leavitt was Clinton's SEC head for eight years, a job that entailed relentless political combat. Leavitt was in his early sixties when he took the SEC job. He had spent his years on Wall Street, but he was the son of a New York State politician and had long been drawn to public service. Little did he know what a blood sport politics had become.

Leavitt's legacy is mixed. He was on the wrong side of important debates over derivatives regulation and accounting rules around stock options. Often he was too cozy with industry lobbyists. But as his tenure went on, Leavitt increasingly targeted the accounting profession for an overhaul. Serious reforms were long past due.

Accountants have been complicit in business crimes as long as there have been accountants. Arthur Andersen and other accounting firms were linked to some of the biggest scandals on Wall

Street in the 1920s. The SEC's security laws, created under FDR in the mid-1930s, were intended to keep the accountants, along with the CEOs, honest. Yet during the bull market of the 1960s, a rash of false earnings statements and other scams revealed widespread dishonesty among accountants and triggered outrage among investors. Members of Congress gave thundering speeches about dirty accountants. A huge congressional study blasted the industry in 1977 and demanded sweeping new regulations to keep accountants honest. Many of these recommendations were not enacted.[11]

Instead, safeguards on accounting grew weaker in the 1980s. The antigovernment zealots of the Reagan Administration wanted to dynamite the SEC, along with every other regulatory agency in Washington. "The other day I was asked if I knew what a 'damn shame' is," commented John Shad, the right-wing Wall Streeter that Reagan picked to head the SEC. "I learned that a damn shame is a busload of government officials going off a cliff—with five empty seats."[12]

Shad's appointment was a classic case of the fox being asked to guard the chicken coop. He set out to gut the SEC in what the *Wall Street Journal* called "the most sweeping deregulation in the agency's 50 years." Shad oversaw the dismantling of stock registration rules, corporate disclosure requirements, and brokerage-house regulations. Even some Republicans were troubled by the abdication of the SEC's role in enforcement. "Wall Street is a cesspool of hanky-panky and the biggest gambling casino in the country," complained a Republican former commissioner in 1982. "The SEC has got to let the industry know that there's a cop on the beat." Shad sent the opposite message.[13]

Shad lost some of his fervor for gutting the SEC as a crime wave engulfed the business world during the mid-1980s. He went on to prosecute Ivan Boesky, Michael Milken, and other insider traders with considerable zeal. In many ways Shad was cleaning up his own mess. Deregulation not only sent a permissive signal to Wall Street, it also led to increased corruption among accountants. While overshadowed by the insider trading cases and big scandals involving defense contractors, the 1980s saw the largest wave of accounting abuses in history. By the early 1990s, accountants had paid out some $1.6 billion in damages in response to over 4,000 liability suits filed in the wake of their role in various scams, most notably the savings and loan scandals. Arthur Andersen paid tens of millions of dollars to settle a suit arising from its role in the failure of the notorious Lincoln Savings and Loan Association. KPMG, Deloitte & Touche, and Ernst & Young were also implicated in S&L collapses. "Accountants didn't cause the S&L crisis," said Senator Ron Wyden, "but they could have saved taxpayers a lot of money if they did their jobs properly and set off enough warning alarms for regulators."[14]

As in the 1970s, the accounting scandals of the 1980s were followed by public recrimination and tough talk from Washington officials—and very little action to prevent future abuses. The first Bush Administration did next to nothing in the early 1990s to tighten rules or toughen penalties in order to keep accountants honest. After all, the administration's goal was to fight "red tape," not add more—even for the purpose of shoring up one of capitalism's key foundations, namely, honest numbers. The stage was set for history to repeat itself.

As the boom of the 1990s gathered steam after Clinton's reelection, signs abounded that accountants were facilitating a new era of business scandals. The SEC's chief accountant under Leavitt, Lynn Turner, later estimated that investors lost as much as $100 billion owing to corrupt audits in the 1990s, even before the meltdown of Enron.[15] In 1997 alone, more than 100 companies filed restatements of their earnings to correct false information. A 1998 survey of CFOs at top companies found that more than half said they had resisted pressures to cook the books in one way or the other; 12 percent admitted falsifying financial data.[16] "The brakes on the worst instincts of the business community weren't working," Leavitt said later. "The gatekeepers were letting down the gates."

Leavitt was particularly worried about the growing conflict-of-interest problems that arose as the big accounting firms increasingly diversified into the non-auditing business. Accountants embraced consulting services in the 1980s and 1990s for a variety of reasons. A lax regulatory environment made it legal for them to take on consulting services for the same clients they audited, while the advent of personal computers decreased their revenues from auditing work. Who needed to bring in a bunch of geeks with green eyeshades when there was Lotus 1-2-3 and Quicken? As traditional accounting revenues shrank, the big firms sought to broaden their relationships with corporate clients. "Because of their training and experience, there was virtually no limit to the consulting services CPAs were asked to provide," observed Philip Chenok, head of the American Institute of Certified Public Accountants during the 1980s and early 1990s. Chenok actively promoted his constituents' credentials as strategic planners, marketing consultants, tax specialists, executive headhunters, lobbyists, financial planners, litigation experts, and even IT specialists and programmers. The accounting industry also labored hard to prove that there was nothing troubling about the consulting craze. As one industry white paper stated, there was "no linkage between providing such services and an impairment of auditor objectivity and independence."[17]

Consulting became the mainstay of Big Five firms such as Arthur Andersen and KPMG, eventually bringing in 70 percent of their revenues. These firms often sought auditing work with a company simply as an entry point for more lucrative consulting deals.

It didn't take a professional ethicist to see how the dual role of watchdog and consultant might lead accountants to turn into spineless yes-men when companies cooked their books. Even as accountants pledged undying allegiance to the GAAP, human nature suggested things could easily play out differently. In a business journal article entitled "The Impossibility of Auditor Independence" and filled with psychological data, two academics observed that auditors might cheat without even realizing it. "When people are called on to make impartial judgments, those judgments are likely to be unconsciously and powerfully biased in a manner that is commensurate with the judge's self-interest," the authors said. Even well-intentioned "auditors will unknowingly misrepresent facts and unknowingly subordinate their judgment to cognitive limitations."[18]

Leavitt's SEC was convinced that this dynamic was playing out across the accounting world. Wall Street wise men like former Fed chief Paul Volcker weighed in with the same view: "this is a clear, evident, growing conflict of interest, given the relative revenues and profits from the consulting practice."[19]

If this problem was easy to identify, a solution was almost impossible to enact in the Washington of the 1990s. Like other industries, accountants had grown increasingly hip to the ways of Washington and put their money where their interests were. Between 1989 and 2000, accounting firms gave nearly $40 million in political contributions. As this money flowed into campaign coffers on both sides of the political aisle, accountants suddenly found themselves with a lot more friends on Capitol Hill. They hired D.C. lobbyists with fat Rolodexes to deepen these relationships, spending upwards of $7 to 8 million a year on lobbying by the late 1990s.[20] They also found natural allies in the phalanx of well-funded conservative think tanks like the CATO Institute and the Heritage Foundation that opposed new regulations on principle.

Leavitt was initially open to the idea that the accounting industry could police itself and create new rules to insure auditor independence. But when the industry proved unwilling to do so, Leavitt sought to force change. The accounting industry responded by upping its spending on lobbying and campaign contributions and by hiring securities lawyer Harvey Pitt as its hatchet man in this fight. (Pitt was later George W. Bush's choice to lead the SEC in a John Shad redux.) One of Leavitt's many adversaries in this battle was Michael Conway of KPMG.

At the time that the battle over new auditor independence rules reached a peak in 2000, Conway was chairman of the SEC Practice Section Executive Committee of the American Institute of Certified Public Accountants. AICPA is the main professional organization of accountants and is responsible for establishing ethics for CPAs, including auditing standards. The Practices Section oversaw self-policing and peer review functions aimed at keeping accountants honest. Both Conway and many of his colleagues at AICPA opposed Leavitt's new rules, which jeopardized the billions of dollars in fees that accounting firms made annually from consulting. They fought Leavitt at every turn.

When Leavitt asked an independent regulatory group, the Public Oversight Board (POB), to investigate the problems with auditor independence, Conway helped orchestrate a brazen move by which AICPA—which provided funding to the POB—tried to kill the investigation by cutting its funds to the POB. The move was darkly ironic, in that one of the POB's mandates was to serve as a watchdog over the AICPA and keep it honest. Conway also tried to head off new government regulations by helping put together new—but largely toothless—measures by which the industry would police itself.[21]

Everyone knows how this story ends. Leavitt didn't win his long battle for tough new rules for auditor independence, settling instead for a watered-down outcome. The bankruptcy of Enron—a company that paid Arthur Andersen half a million dollars a week in consulting fees while also having it police its books—was just one of the many scandals related to cooked books and rotten accountants. Compromised accountants have also been implicated in the scandals at WorldCom, Tyco, Global Crossing, and many other companies. Millions of Americans would have more money in their retirement accounts today, more money saved for their children's education—more peace of mind and more security—if the accounting industry had been forced to clean up its act. "If there ever was an example where money and lobbying damaged the pubic interest," Arthur Leavitt said later, "this was it."[22]

The failure to ensure the independence of auditors was only one of three major regulatory failures that led to the corporate scandals. A second was the fall of the "Chinese Wall" that was supposed to separate stock research analysts from investment bankers, providing the incentive for star analysts like Henry Blodget and Jack Grubman to mislead investors on a massive scale. These analysts publicly rated the stocks of companies that also paid banking fees to the firms where they worked. In principle, analysts are supposed to spend their days buried in company earnings reports and economic data, emerging bleary-eyed from their offices to issue critical and independent judgments about stock values. They are supposed to be strictly walled off from the greedy bankers by unbreachable institutional barriers. Instead, many analysts spent the boom of the '90s neck-deep in the investment-banking activities of their firms. Analysts didn't just talk to the bankers; they were commonly compensated for bringing investment clients to the firms. When the "analysts" went on to publicly dissect the prospects of these companies, sometimes on television shows watched by millions of investors, they failed to disclose their massive conflicts of interest. It was an appalling game. Wall Street insiders knew all about the game and took everything that conflicted analysts said with a grain of salt. But most Main Street investors didn't have a clue.

In 2003, government regulators led by New York State attorney general Eliot Spitzer reached an agreement with some of America's top financial-services firms to stop these practices and rebuild the Chinese Wall. The question, of course, is why this didn't happen earlier. The answer is that regulators didn't have the resources or political muscle to make it happen.[23]

A third regulatory failure in the '90s was the inability of regulators to compel corporations to report stock options as expenses in their earnings reports. This not only gave corporations more incentive to dole out huge numbers of options to top executives—thus providing temptations to concoct false earnings statements and boost stock prices—it also allowed many companies to report higher profits than they actually had, since one of their biggest personnel outlays was not formally listed as an expense.

A number of government officials in Washington, most notably Michigan senator Carl Levin, pressed in the early 1990s to have stock options categorized as expenses. Some business leaders like Warren Buffett also backed this idea. The SEC even took tentative steps in this direction. But a revolt against the move quickly gained steam, led by the high-tech industry, which relied heavily on options to compensate its employees. Democratic senators like Joseph Lieberman and Dianne Feinstein lined up with industry lobbyists to successfully block the proposal. Commonsense regulation suffered yet another setback.

As the boom of the '90s got going, a deeply corrupt system was in place whereby executives had huge temptations to cheat on their earnings statements, accountants had huge temptations to sign off on these bogus statements, and top stock analysts had huge temptations to exaggerate the value of stocks that investors were buying. Respected CEOs, accountants, and stock analysts gave in to these temptations not because they were uniquely evil and greedy but because the rules of the system allowed it. In a very short time, as the '90s boom reached a crescendo, that system enabled a small elite of insiders in corporate America and on Wall Street to become unbelievably wealthy—while ordinary Americans picked up the tab.

TAXES ARE ANOTHER area where government enforcers were disarmed during the '90s and new temptations to cheat arose. Nobody knows this better than the people at the IRS.

Consider the vantage point of a top IRS official in 2001. During the months leading up to tax day, which fell on April 16, the IRS received tax returns for the biggest boom year in U.S. history. So much money was rolling into the Treasury that budget experts forecast trillions of dollars in surpluses over the next decade and beyond. But instead of celebrating, IRS officials were gnashing their teeth about all the money that got away—which was estimated to be over $250 billion, an amount that could have covered most of the Pentagon's budget that year.

The IRS has a formula known as the "tax gap" that is used for estimating revenues lost to tax evasion. The tax gap compares the money being earned by individuals and businesses to the money that gets paid in taxes. Over the past twenty years, the tax gap has soared.

Back in 1990, the IRS estimated the tax gap at $100 billion. Noting that tax evasion had increased in the 1980s, the IRS suggested that the likely cumulative value of tax avoidance since 1972—$1.6 trillion—was roughly equivalent to 60 percent of the 1990 federal debt. In 1998, IRS officials submitted their highest-ever estimate of the so-called tax gap—$195 billion annually. The figure represented an increase of nearly 100 percent since 1990. And it kept climbing. The IRS estimated the tax gap for 2001 at $250 billion. Outside experts put the annual total much higher.

According to former commissioner Donald Alexander, the gap between what taxpayers owe and what the IRS collects could be as high as $500 billion a year. This staggering figure, it should be said, does not include evasion of state and local taxes.[24]

The IRS knew full well that the government wasn't getting close to what it deserved in 2001. But it had few options for dealing with this problem, since for over twenty years the agency has been underfunded and undermined by Congress and the White House.

The first problem the IRS faced after April 16, 2001, was just trying to figure out the basics of who might be cheating. Stone Age technology and a lack of staff made the task nearly impossible. The central computer system at the IRS, built in the 1960s and maintained by a series of adjustments and improvements in the intervening years, relied on magnetic tapes to keep track of the records of the nation's taxpayers. By the late 1990s, the system ranged across 147 different mainframes and employed 8,700 different software products. Testifying before Congress in 1999, IRS commissioner Charles Rossotti explained that the only reason IRS computer databases remained accessible was that longtime employees knew how to jerry-rig the system to produce results.[25]

Even with the best technology, the IRS would have been in a hopeless situation since it was understaffed in 2001 and remains so today. Between 1989 and 1999, the total number of IRS employees fell by 26 percent. The staff of the IRS Office of Examination, which includes tax auditors and revenue agents, fell by 34 percent during this period. And yet the number of tax returns filed by individuals increased by 14 percent in the same decade. Many people's tax returns also became more complex. The number of taxpayers claiming $100,000 or more in income quadrupled during the 1990s boom, with many filers itemizing numerous deductions. The ranks of self-employed taxpayers rose dramatically as well—26 percent between 1988 and 1998, or twice the rate of those reporting wage income.[26]

Also challenging was the proliferation of tax returns with income from so-called K-1 partnerships, which include several different business arrangements. By 2001, tax returns were filed for around 13 million K-1 partnerships. This income is particularly difficult to verify because while employers file W-2s for wage earners and 1099 forms are typically filed for consultants and other kinds of self-employed people, there are often no third parties vouching for the income paid through K-1 partnerships. This leaves the IRS to laboriously track such income through other means.

Complex tax-shelter schemes proliferated during the 1990s, many of them offered by respected banks and brokerage firms that previously had steered clear of this shady practice. And the tax shelters used by corporations were even more byzantine than those used by individuals. A survey by the Institute for Taxation and Economic Policy found that 52 of the 250 largest U.S. companies paid taxes at a rate of 10 percent in 1998 even though the overall tax rate for corporations in the top ranks was 35 percent. In 1999, corporate profits surged 8.9 percent, even as corporations paid 2.1 percent less in income tax than they had the previous year. The use of illegal tax shelters by corporations was thought to cost the government about $10 billion a year in 2002.[27]

By the late 1990s, the IRS's enforcement powers had become a joke. Its capacity for tracking and verifying income has been stretched as more Americans have come to play more sophisticated games with their money. In turn, the perception of a strapped IRS has led many Americans to think, correctly, that they can get away with tax evasion. Investigations and prosecutions by the IRS fell off sharply during the 1990s, dropping by half even as more returns were filed and cheating increased. In 2002, the IRS was aware of nearly 2.5 million individuals who had failed to file federal income tax returns. A surprising number of these nonfilers included affluent professionals, such as doctors, lawyers, and other self-employed professionals. Due to budget constraints, 75 percent of these cases would not be investigated.[28]

Though not everyone has benefited from the IRS's new tolerance for tax evasion—wealthy taxpayers certainly have. The wealthier you are, in fact, the less the IRS bothers you. Since 1988, audit rates have increased by nearly a third for poor tax filers—while dropping 90 percent for the most affluent earners. In 2001, poorer individuals making under $30,000 who made use of the Earned Income Tax Credit had a 1 in 47 chance of being audited. Meanwhile, taxpayers making over $100,000 had a 1 in 145 chance of being audited. Lawyers, doctors, and other professionals with S corporations had a 1 in 233 chance of being audited. Those with partnerships had a 1 in 400 chance of having IRS agents show up at their door. Not surprisingly, wealthy Americans express much less concern about being audited than their lower-income counterparts. In a 2001 poll, 18.9 percent of taxpayers earning more than $100,000 reported worrying about audits, as compared to 32.7 percent of those earning less than $25,000.[29] "It's a risk-reward assessment," says Gail Kenney, the IRS's director of financial crimes. "If the likelihood is they won't get caught, and that, if they do, there won't be serious consequences, then they'll do it."

The IRS goes after less-affluent tax cheats because they are easy prey. Middle-class and lower-income Americans don't have the resources to put up a lengthy fight against the IRS. But America's wealthiest taxpayers, backed by top tax lawyers and accountants, are adept at tying IRS investigators up in knots for years. A 2001 investigation by the Center for Public Integrity, a Washington research group, observed that "hundreds of the best tax minds in the nation, including former [IRS] commissioners and other ex-IRS officials, reap millions of dollars annually by helping the largest corporations and the wealthiest citizens avoid paying their fair share of taxes." In numerous cases in the '90s, the IRS went after wealthy tax cheats seeking millions of dollars in unpaid taxes—only to end up with a fraction of this after years of investigation and negotiation. As one IRS lawyer has commented: "When there are ten thousand documents, some of which are bank statements containing thousands of transactions, and the opposition argues over the significance of every single item, the process becomes extraordinarily difficult." In the face of battles like these, it simply makes more sense for the IRS to concentrate its enforcement efforts on vulnerable targets.[30]

The IRS's enforcement priorities have long been well-known among the wealthy. In Leona Helmsley's 1989 trial for tax evasion, Helmsley's head housekeeper at her twenty-eight-room mansion in Greenwich, Connecticut—where a variety of improvements had been improperly claimed as tax deductions—testified on the stand about an exchange she had with the wealthy hotelier. "You must pay a lot of taxes," the housekeeper said. Helmsley replied, "We don't pay taxes, only the little people pay taxes."

During the 1990s, judicial and legislative actions made it even easier for the rich to thumb their noses at the IRS. A 1991 Supreme Court ruling opened the door to more widespread tax avoidance by allowing tax evaders to exonerate themselves by citing ignorance of the law and forgetfulness. Under the principles outlined in the *Cheek v. United States* decision, prosecutors now bear the heavy burden of proving that individuals "willfully" violated the law.[31] And Congress's hostility toward the IRS has been legendary. In 1998, the Gingrich Congress passed the Taxpayer Bill of Rights, which limited the ways that the IRS could pursue tax cheats. Although the law curbed some actual abuses, the real agenda of House Republicans was to weaken the IRS. Money from corporations and wealthy individuals funded a "grassroots" campaign to help pass the law. In a rare public unmasking of the right's tendency to disguise plutocratic goals in populist clothing, many of the "average people" who testified before Congress about IRS abuses were found to be phonies. According to a report in the *Christian Science Monitor,* most of the IRS sob stories turned out to be patently false.[32]

The Taxpayer Bill of Rights was a huge victory for all tax cheats, but its main beneficiaries are taxpayers wealthy enough to afford accountants or lawyers who know the ins and outs of the law, and can exploit its provisions to make the IRS back off. The year after the law went into effect, tax advisers were coaching clients on how to make the best use of its provisions to reduce tax burdens and postpone the payment of outstanding taxes. "These measures will diminish compliance with the tax law and the honest taxpayer will end up even more disgusted," complained former commissioner Donald Alexander.[33]

The corporate scandals focused new attention on the problem of tax evasion, especially among CEOs and other wealthy Americans who were making widespread use of dubious tax shelters. The 2001 Bush Administration offered some response to this problem by modestly increasing the IRS's budget. However, when departing IRS commissioner Charles Rossotti prepared his final report to Congress late in 2002, calling the budget increases grossly inadequate for deterring fraud, the Bush Administration successfully censored his remarks.[34]

HOW IS IT that wealthy insiders have been able to hijack America's financial system and tax system for their own gain at the expense of everyone else? Much of the answer lies in the outsized political clout that economic winners have in our society today.

It is no secret that income gaps translate into unequal political muscle. Politicians pay attention to voters, but middle-class and low-income Americans are less likely to vote than those on the upper rungs of the economic ladder. Politicians pay attention to campaign contributors, but ordinary working Americans are far less likely to make political contributions than upper-income people. Politicians pay attention to citizens who write them letters or call them or show up at public hearings or attend protests or sign petitions or volunteer for a political campaign. But wealthier Americans do all of these things at higher rates than other citizens.[35]

Education is part of the problem. Less-wealthy Americans tend to be less educated and are less likely to keep up with politics or feel confident navigating their way through civic processes. Many of these Americans are also convinced that their vote and voice does not matter, that the system is rigged in favor of those at the top, and that the little people don't count. This outlook can be a self-fulfilling prophecy. The more that ordinary Americans get cynical and tune out of politics, the more likely it is that politicians will ignore their interests altogether. The squeaky wheel gets the grease in politics, and there's not much incentive to stand up for constituents who don't stand up for themselves. Yet even when average citizens do stand up for themselves—even when they band together in large numbers—it's hard to be as squeaky as special interest groups, with their fat-cat lobbyists and checkbooks open to politicians.

That the rich wield a megaphone in American politics is nothing new. But in the past two decades, not only have the rich gotten dramatically richer compared to everyone else, they have found many new ways to leverage that wealth in the political sphere. By influencing laws and regulations at every level of government, they use their wealth to bend rules to allow cheating, rewrite rules to redefine cheating as legitimate behavior, and walk away unpunished when they are caught cheating.

Consider the swollen river of money that now flows into our politics. Most Americans are mainly familiar with the cash that goes to political candidates and parties. This spending has increased dramatically in the past two decades. During the 2000 election, $4 billion was spent in all of the federal and state elections—some three times what was spent in 1992. Much of this money was in the form of very focused campaign gifts by industry and the number of corporate political action committees has doubled since 1980. For example, the $40 million that the accounting industry pumped into politics during the 1990s didn't go to just any politicians; it went to politicians who sat on key committees or otherwise were in a position to do favors for accountants. (A total of $640,000 also went to George W. Bush's 2000 campaign.) Ditto for the million that the pharmaceutical industry gave during this same period, or the real estate, oil and gas, and telecom industries. The favors done by politicians in return for this money underscore that writing checks is the form of political engagement that matters most these days—one that is beyond the means of the tens of millions of ordinary Americans who have more pressing checks to write every month.

Often, the sleazier a company is, the more it pays politicians to either look the other way or pass rules that permit bad behavior. Two of the biggest political contributors in the 2000 election cycle were Enron and MBNA. Enron is now well known to have been a cesspool of cheating. MBNA, the nation's largest credit card company, has no such reputation. Why? Because many of the blatantly unethical practices of this and other major credit card companies—including deceptive advertising, usurious rates, hidden fees, and excessive penalties—have been legalized over the past quarter century through a deregulatory process lubricated by political contributions. MBNA's heavy investment in politicians of both parties is aimed at keeping the favors coming.

Contributions to political campaigns are only one tributary feeding the mighty river of money in politics. Another is the growing influx of money into think tanks and advocacy groups.

The American Enterprise Institute is a good example of intellectuals for sale. Lodged in a modern office building in downtown Washington, AEI is the premier scholarly institution of the conservative movement. Founded in 1943, it initially came to distinguish itself over decades for its genuine commitment to advanced scholarship, albeit with a moderate rightward tilt. As it happened, though, that tilt was not nearly right enough for those who controlled AEI's purse strings. In the mid-1980s, AEI's main contributors threatened to cut off funding unless it turned much more conservative.

It did. These days, AEI has a $17 million budget and provides a home to some of America's most rabid right-wing ideologues, including Robert Bork, Newt Gingrich, and Charles Murray. AEI's twenty-five-member board of directors includes exactly one scholar, James Q. Wilson. The rest of the board is packed with corporate CEOs, including the chiefs of ExxonMobil, Dow Chemical, State Farm Insurance, and American Express. These extremely busy men aren't there because they enjoy a good intellectual conversation. They are on AEI's board because their companies are among the dozens that donate handsomely to AEI, funding a steady stream of highbrow studies that trash government regulation, advocate repealing taxes on corporations and the rich, propose ways to dismantle America's social safety net—and even seek to rehabilitate social Darwinist ideas about the innate superiority of some groups of human beings over others, as AEI did when it supported Charles Murray's research for his controversial book on human intelligence, *The Bell Curve*. "Corporations provide important input to AEI on a wide variety of issues," admits AEI's annual report. Yet what serious think tank would want input from entities designed solely to maximize shareholder value? Self-interest is why so many corporations give money to AEI—over $5 million a year—but self-interest is antithetical to what sound scholarship is all about.

Even as the American Enterprise Institute has tacked to the right over the past two decades, it remains known as one of the more reasonable conservative think tanks. A number of well-respected scholars still reside at AEI. The Heritage Foundation, with nearly twice the annual budget of AEI, has no such reputation. Heritage is a factory for far right-wing ideas and a virtual arm of the GOP. It is heavily funded by wealthy conservative zealots, including the reclusive Richard Mellon Scaife, who also bankrolled a $2 million smear campaign against Bill Clinton. (All contributions to Heritage, it should be added, are tax deductible, because it is a nonprofit organization.) Heritage's researchers grind out reports with titles like "How to Close Down the Department of Labor." While Heritage's research is widely criticized as biased and shoddy, other conservative think tanks suffer far worse credibility problems, at least among those in the know. The Employment Policies Institute, for example, exists almost strictly as a creature of the restaurant industry. And guess what? EPI's research shows that raising the minimum wage is sure to trigger another Great Depression.

In all, the labyrinth of national and state conservative think tanks funded by wealthy interests now spends nearly $200 million a year to influence public policy. Social science, long a central tool of pragmatic reformers who first objectively studied problems and then developed solutions to them, is increasingly used as a weapon by wealthy elites to get their way. The media, committed to telling both sides of any story, play right into this strategy by placing the views of independent experts and corporate-funded intellectuals on an equal plane. An eminent university expert on pension systems will be quoted as saying that privatizing Social Security would hurt low-income seniors—and be contradicted a paragraph later by a "resident scholar" whose think-tank paycheck comes indirectly from a financial services industry that dreams of 150 million private accounts. A Nobel Prize–winning climate expert will be quoted as saying that global warming would have devastating consequences—only to be dismissed on *Nightline* by a "senior fellow" whose work is funded by an oil industry that couldn't care less if Florida ends up under water in a hundred years. The public, which is prone to tune out whenever policy wonks start yammering on anyway, now has an even harder time sorting out objective research findings from sophisticated propaganda.[36]

Wealthy interests have also gotten much better in recent years at buying faux grassroots support. Take an outfit like Citizens for a Sound Economy, a deceptively named organization if ever there was one. Although it boasts 280,000 citizen members, CSE is purely a top-down creation of major corporations and rich donors who underwrite its operations. From headquarters in Washington, CSE specializes in orchestrating campaigns that make elected leaders think there's a groundswell of grassroots support for a given policy position when, in fact, there is not.

If 150,000 faxes and e-mails arrive in congressional offices all advocating a repeal of the "death tax" because it hurts small farmers and businesses, you can bet that CSE is behind it. (In fact, the estate tax is paid almost exclusively by multimillionaires, although years of deceptive analysis have now convinced the public otherwise.) When protestors picket the offices of the Food and Drug Administration, complaining that its lengthy drug review process is harming sick children, CSE and its backers in the pharmaceutical industry are likely pulling the strings. CSE also uses television advertising and direct mail to blast away at politicians in their home districts who are unfriendly to corporate interests. One of CSE's biggest backers is the Charles G. Koch Charitable Foundation, as in Koch Industries, the energy company that paid $40 million in fines in 2000 and 2001 alone for breaking environmental laws.[37] In recent years, CSE's success has been increasingly copied by other organizations, and there is now an impressive constellation of groups in Washington that purport to speak for ordinary Americans but actually speak for the very richest Americans. My favorite these days is the Seniors Coalition, a Beltway group that says it "represents the interests and concerns of America's senior citizens at both the state and federal levels"—but which in fact is funded mainly by corporate interests to push a Republican free-market agenda, including the privatization of Social Security.

Neither politicians nor intellectuals nor faux citizen activists come cheap. Yet the huge aggregation of wealth at the top of the income ladder means that there's been no shortage of cash to pump into politics. And even as the two tributaries of campaign monies and funds for think tanks and advocacy groups have grown rapidly, a third tributary in the river of political money has overrun its banks—money for lobbying. Today, Washington is swarming with hired guns as never before. Money spent on lobbying climbed to $1.5 billion in 1999 (the last year for which there is current data) and Washington now has nearly forty lobbyists for every member of Congress. Many sleepy state capitals have also been overrun by the well-coiffed representatives of business.[38]

It would be one thing if all these lobbyists did nothing more than work the halls of Congress or the state legislatures, where a growing number of them once worked before they decided to make real money. The reality is that lobbyists do much more than that. They also actively participate in drafting legislation and regulations. This has been particularly true under the George W. Bush

Administration, where industry representatives have been asked to participate in various tasks forces and internal commissions. In fairness to Bush and the Republicans, hands-on industry involvement in regulatory decisions is hardly new. It's just become deeper and more pervasive.

In today's Washington, those in power never forget whom they serve. And while the co-opting of America's political leadership by wealthy interests helps explain why the SEC, the IRS, and many other watchdog agencies don't stand up to wrongdoers, it also explains why conservatives get away with framing America's debate on values so narrowly. Even as conservative politicians talk endlessly about the temptations that come from illegal drugs or sex education or no-fault divorce laws, there are far fewer politicians ready to discuss the moral hazards of an unregulated market.

After all, they have reelection campaigns to fund.

CHAPTER SIX

Trickle-down Corruption

FOR THOSE WHO ARE PART OF THE WINNING CLASS, OR TRYing to be, there are plenty of reasons to cheat. The rewards are bigger and the rules are toothless. Yet many Americans with more modest ambitions and more humble means are also cheating.

Take the mild-mannered bookkeeper as an example. He is, by all appearances, an honorable man. He neither drinks nor smokes, and is quiet and dependable in the way of many bookkeepers. He rarely misses a day of work or tarries on his lunch break. When United Way comes around, he always contributes. He and his wife lead an orderly life with their two polite children and spend Sunday mornings at church.

The bookkeeper works hard during the early years of his job and finally gets up the gumption to ask for a raise of $100 a month. He is crushed when the request is denied. But the bookkeeper seems to get over his disappointment and soldiers on. He still arrives punctually every day. He never calls in sick when, in truth, he is well.

After twenty years, the bookkeeper finally retires. The company throws a small farewell party for him and gives him a watch.

He and his wife pack up for Florida to start their golden years. A new bookkeeper takes his place. Poring over the financial records, this new bookkeeper finds that something is wrong. Things aren't adding up. He flags his concern to the company. No, no, he's told, the old bookkeeper would never get into any fishy business. He was a rock of reliability, the soul of integrity.

And yet, when the new bookkeeper completes his investigation, the facts are incontrovertible. The old bookkeeper, it is clear, engaged in a systematic pattern of embezzlement. The pattern is oddly consistent. Year after year, the amount of money stolen is never greater and never smaller, nor is it particularly large. It is $100 a month.

The thieving bookkeeper exists in an apocryphal story passed down over many years among fraud examiners who probe workplace theft. The story is told to illustrate a point these investigators know all too well: that people are prone to invent their own morality when the rules don't seem fair to them. This tendency explains a lot of cheating in America today.

There are roughly four reasons why people obey rules. First, we may toe the line because the risks of breaking the rules outweigh the benefits. Second, we might be sensitive to social norms, or peer pressure—we follow the rules because we don't want to be treated as a pariah. Third, we may obey rules because they agree with our personal morality. And fourth, we may obey rules because they have legitimacy in our eyes—because we feel that the authority making and enforcing the laws is just and ultimately working in our long-term interests.

When people don't obey the rules, you'll often find several things going on at once. The Winning Class cheats so much because there's more to be gained nowadays and there are fewer penalties, either legally or socially. Students often cheat for the same reasons: the stakes of academic competition are higher and the normalization of cheating means that there's little peer pressure to be honest.

Motives like these are not hard to understand. Cases like the bookkeeper are more complex. A simple risk/benefit analysis doesn't explain everything, since the bookkeeper was running a serious risk for only a modest sum of money and could easily have taken more. Nor do social norms offer much insight, since the bookkeeper's thefts were not condoned by his peers. Instead, the bookkeeper operated by his own moral code to take from the company what he felt it owed him.

A lot of Americans have been inventing their own morality lately. Tens of millions of ordinary middle-class Americans routinely commit serious crimes ranging from tax evasion (a felony), to auto insurance fraud (also a felony), to cable television theft (yes, a felony as well in some states), to Internet piracy of music and software (more felonies). Most of these types of crimes are committed for small potatoes: to receive $700 more on a tax refund, to save $400 a year on an insurance premium, to get $40 a month worth of premium cable or an $18 CD for nothing. These crimes are being perpetrated by people who see themselves as law-abiding citizens, people who don't imagine themselves above the law and who don't have big-shot lawyers on call.

Day-to-day criminality among ordinary Americans is nothing new. "Unlawful behavior, far from being an abnormal social or psychological manifestation, is in truth, a very common phenomenon," commented the authors of a 1947 article about "law-abiding law-breakers."[1] However, evidence indicates that this familiar problem has worsened in recent years—even as conventional street crime has fallen dramatically.

What is going on here?

Much of the answer, I suspect, lies in our broken social contract. An orderly democratic society depends on having a social contract in place that delineates people's rights and responsibilities. It also depends on people having faith that the social contract applies fairly across the board. The social contract will break down when those who play by the rules feel mistreated, and those who break the rules get rewarded—which has been happening constantly in recent years.

John Q. Public need not to be versed in John Locke to feel that he has a legitimate cause for cynicism. He knows that white-collar criminals walk free, that fat-cat tax cheats get off the hook, that corporate money buys political favors, and that Ivy League schools are filled with kids whose rich parents greased the system to get them in. He also knows that when there's a war, it's working-class kids who fight it; when there's a tax cut, he probably won't get more than peanuts; and when there are layoffs, it's those lower down the totem pole who'll get the ax.

Polls confirm that many Americans see "the system" as rigged against them. When asked who runs the country, many say corporations and special interests. When asked who benefits from the tax system, most say the rich. When asked who is underpaid in our society, most agree that lots of people are underpaid: nurses, policemen, schoolteachers, factory workers, restaurant workers, secretaries. And when people are asked whether it is possible to get ahead just by working hard and playing by the rules, many say that it is not.[2]

It is commonly said that Americans tolerate huge income gaps and other unfair advantages of the rich because they think that they themselves will be rich someday. But polls show that most people don't actually see great wealth in their future. Less than half of Americans think they will ever be rich, and up to half believe that their children will be worse off than they are. At the same time, nearly half of Americans surveyed in 2001 said they saw U.S. society as divided between haves and have-nots, compared to about a quarter of Americans who felt that way in 1988. More startling, after years of prosperity, is that the percentage of Americans who identify themselves as being among the have-nots has nearly doubled, to a third, since then.[3]

As wealth and power inequities have grown in America—as the rich have grabbed most of the new wealth and come to live in their own moral community with its own set of rules—ordinary Americans have taken note. Many Americans see plenty of evidence of a broken system in their own economic struggles. Polling shows that as many as 50 percent of Americans worry that they will be poor at some point in their lives. Extensive economic data that compares wages with the cost of living in America shows that over a third of all households do not earn enough income to meet basic needs. Opinion polling confirms that a third of families consistently report that they are "pinched" or not making ends meet.[4]

Economic insecurity is nothing new to the bottom third of American households. But the downsizing phenomenon of the '80s and '90s has also meant less job security for professionals who should reside securely within the middle class. Most of those who are downsized find other jobs, but these jobs often pay less and carry few pension or health-care benefits. Those who survive downsizing find themselves expected to work harder and produce more, although often with greater pay. "Work in the past 20 years has grown more insecure," conclude Neil Fligstein and Taek-Jin Shin, two scholars who examined trends between 1976 and 2000. "Job tenure is down for everyone and the possibility that workers will have to take temporary work or work involuntarily has risen.... The changes in the security of work were mirrored by changes in benefits and health and safety at work. Over time, health and pension benefits decreased for all workers."[5]

None of this means that Americans are ready for a socialist revolution. As David Brooks and a thousand other social critics have pointed out, Americans vote their dreams, not their realities. The middle class and the working class consistently elect politicians who actively work against their economic interests. Perhaps at no time in American history is this more true than today.[6]

The psychological fallout from people's economic struggles has been significant. People worry intensely about their finances, especially the heavy debt burdens that they often carry.[7] Many people are also less happy. "Happiness and satisfaction with life are, in many ways, the ultimate bottom line, a test of the good society," observes scholar Michael Hout. Yet in the past quarter century, Hout's work shows, gains in happiness have not been shared evenly in a U.S. society more divided by income: "the affluent are getting slightly happier and the poor are getting sadder; the affluent are increasingly satisfied with their financial and work situation while the poor are increasingly dissatisfied with theirs."[8]

Such endemic unease might itself be a corrupting force in society. But economic struggle is all the more dangerous when mixed with high expectations of well-being—that is, the expectation that one that should be as happy as the shiny rich people on television and in magazines seem to be. Writing in the mid-twentieth century, the sociologist Robert Merton observed that Americans are taught that everyone can succeed if they work hard enough. America was "a society which places a high premium on economic affluence and social ascent for all its members." But Merton also pointed out that there is no "corresponding emphasis upon the legitimate avenues on which to march toward this goal." Americans worshipped financial success without being too concerned about how people got ahead. "The moral mandate to achieve success thus exerts pressures to succeed, by fair means if possible and by foul means if necessary." These pressures were especially poisonous, Merton said, in a nation where not everyone actually could succeed—where there were limits on the economic opportunities that were available.[9]

Merton could have made these points yesterday. The pressures on Americans to make a lot of money are extremely high—higher, maybe, than they've ever been before. To be sure, there are many legitimate opportunities to do well financially. Yet ultimately the opportunities are finite. America needs only so many skilled and well-paid professionals. In an economy where structural conditions allow only the top fifth or so of earners to really get ahead, the other four fifths of Americans are stuck in the bind that Robert Merton identified: they live in a society with insanely high material expectations but with limited ways to meet these expectations.

What to do in this conundrum? Whatever you can get away with.

And how do ordinary, moral people justify doing wrong to do well? Often, they point to the unfairness around them—to the structures that keep them struggling while others thrive, to the ways that bad guys easily climb to the top, to the cheating that goes on by the rich and powerful every day. "People in subordinate positions make moral judgments about existing social arrangements and assert their prerogatives to personal entitlements and autonomy," writes Elliot Turiel, a leading authority on moral development. Turiel is fascinated with why people break rules, and much of his analysis centers on what he dryly calls "asymmetrical reciprocity implicit in differential distribution of power and powers"—in other words, feelings of injustice. Turiel observes that "in daily life people engage in covert acts of subterfuge and subversion aimed at circumventing norms and practices judged unfair, oppressive, or too restrictive of personal choices." These acts may place people on the wrong side of the law, or the established rules, Turiel says, but their true ethical implications are often a fuzzier question. "In my view, it would be inaccurate to attribute these types of acts of deception to failures of character or morality. Many who engage in these acts are people who generally consider themselves and are considered by others as responsible, trustworthy, upstanding members of our culture."[10]

It is easy to cheat like crazy and yet maintain respect for yourself in a society with pervasive corruption. It's easy, for example, to justify cheating in a country like Brazil where oligarchical families have been abusing the little people for a couple of hundred years and are still doing it, or a country like Pakistan where government ministers and their pals in business live in luxury while millions rot in the slums of Karachi.

And more and more, similar rationalizations can work just fine in the United States.

The social theorist Max Weber was among the first scholars to explore how people's views of "legitimacy" shape their respect for rules. He argued the commonsense point that people are more likely to follow rules or laws that seem fair and are made by an authority that deserves its power. There was nothing actually pathbreaking about this point when Weber made it a century ago. Numerous big thinkers going back to Plato had made similar arguments, and support for this idea cut across fields—from political science to anthropology to sociology to education. Yet if the idea seemed like common sense, what Weber and other scholars typically lacked was the empirical "proof." How can you really tell why people either obey the law or break it? How can you weigh legitimacy as a factor when there are so many other influences on people's behavior? I may speed for many reasons: because I'm late or I'm a thrill seeker or I think it's wrong for the federal government to impose speed limits and usurp local authority on this issue. Short of hearing me and lots of other speeders out and somehow verifying that we're telling the truth about our motives, who can say why people like me drive so fast?

Proof that views about legitimacy explain ethical decisions remains hard to come by. But the evidence has gotten a lot more compelling since Weber's day. In his 1990 book, *Why People Obey the Law,* Tom Tyler picked up the legitimacy baton and ran with it into new empirical terrain. Tyler marshaled data going back thirty-five years in arguing that most people are inclined to obey the law, but that this reflex can easily be undermined if the law is widely seen as lacking legitimacy. He looked at studies of juvenile delinquents in England, college students in Kentucky, middle-class workers in Germany, and poor black men in Newark, among others. He also conducted his own large surveys of Chicago residents. Tyler's conclusion after all of this? Pretty much what Weber said a hundred years earlier.[11]

Yet if the link between respecting authority and following the rules has found more support in general, this is still complex terrain. Much of the time when people break rules you'll find a sticky wicket of conflicting evidence about their motives and no easy way to nail down what they were really thinking. Most people don't like to talk openly about cutting corners. Also, the root causes of why people break rules can be obscured when cheating becomes so routine that people no longer give it much thought.[12]

THE CANDOR OF Jennifer Bennett (not her real name) sheds some light on what is going on in many American households—and, in particular, how cynicism and anger might cause a person who normally wouldn't even run a red light to commit a felony that is punishable by up to five years in prison.

Bennett should be one of the good guys in my story. She was raised in New Jersey by parents who taught her to play by the rules. She works in the arts in New York City but is obsessed with neither money nor status. She just wants to do her art and get by. She believes that government can make a difference in people's lives and, if anything, that taxes should probably be higher than they are.

Yet every year, come April 15, she submits a work of fiction to the IRS.

"Much of the money I earn is off the books—it's money earned in cash through private teaching or tutoring. I generally claim a portion of this money, but not all of it," Bennett says. "It's the money I earn to support my pursuit of a career in the arts. I put thousands of dollars a year into this career, pay my own insurance, and receive no benefits. I guess that's the way I justify writing off as much as I can and claiming as little as I can. I feel that most other first-world countries support their artists and the arts in a much stronger way than we do, and that the wealthy in this country are the ones with the real benefits."

Bennett has struggled financially for years, despite her Ivy League degree. Meeting the rent has often been an adventure, and she now lives 130 blocks north of Times Square, in a low-income neighborhood near the George Washington Bridge. "When I see people getting million-dollar bonuses for moving money around, who then walk in free to city museums because their companies are corporate sponsors, my jaw drops. Most artists I know can't afford to attend arts events on a regular basis. I figure the amount of money I earn is so tiny compared to what most people in this city are earning, and that if I had to pay thousands of dollars in taxes at the end of the year, on the relatively small amount I earn, I couldn't afford to continue doing what I'm doing."

Bennett has anguished about her tax cheating—for, like, three seconds—over the past five years. "I don't think it's the 'right' thing to do, but personally, I don't really care. I know that one wrong doesn't right another wrong, but until I see any sign of a real move to universal health-care coverage or the closing of loopholes for the rich, or increased benefits for those making their living in the arts, I don't feel particularly inclined to be honest. When I read about the IRS going after those in lower-income brackets, it makes my blood boil."

Another tax cheat, a man who drove a cab in New Jersey for many years while putting himself through school, says much the same thing. "I hardly reported any of my income. I was barely making enough money as it was and there was no way I was going to hand it over to a government that doles out tax breaks to the rich and leaves students like me to starve." Again and again, tax cheats say the same thing. "I am honest in every other aspect of my life," says a man who claims false charitable deductions every year, a common practice. "I rationalize it by saying that wealthy people and big businesses have much larger loopholes that they're taking advantage of, so that it is very low down on the scale of crimes. The only guilty feeling I have, is the fear of getting caught."

These attitudes toward cheating on taxes are typical. A 2001 poll found that 43 percent of respondents indicated they had cheated, that they would cheat, or that they thought it was okay to cheat. Although other polls indicate that the vast majority of Americans think that it's wrong for anyone to cheat on their taxes, acceptance of tax cheating appears to be on the rise.[13] Estimates of lost revenues by the IRS underscore the widespread nature of tax evasion. As we saw earlier, the most egregious tax cheats can be found among the wealthy—those savvy enough to know that they can get away with thumbing their nose at the IRS—but widespread cheating by more ordinary Americans may make up as much as half of the $250 billion lost every year to tax evasion.

Polls show that Americans have little faith in tax fairness. Between a third and a half of Americans think that the taxes they pay are not fair. More than half said in 2003 that high-income families didn't pay their fair share, while middle-income families paid more than their fair share. Half of respondents also said that the single thing that bothered them most about the tax system—more than their own tax burden or the complexity of taxes—was that "some wealthy people get away not paying their fair share."[14]

Accountants hear such rationalizations nearly every day. "The perception that Congress is intentionally allowing wealthy taxpayers and big business to escape taxes legally can be a strong motivation for others to create their own [illegal] tax savings," the leader of the American Institute of Certified Public Accountants once remarked to a congressional panel. "Some perceive that the IRS enforcement practices are applied in an uneven and inequitable fashion," he explained, "whereby low- and middle-income taxpayers are harassed over small amounts, while insufficient attention is paid to the wealthy."[15]

Other Americans hate taxes for the simple reason that they hate government. Antitax zealotry has risen over the past two decades as we've been repeatedly told that the free market can solve all of our ills and that government is a waste of money or, worse, a plot to deprive us of our rightful private property. If government is the problem, not the solution, why give it a dime? And if government no longer represents "ordinary" Americans, why let it grab a good chunk of our income every year? "There is no difference between an income tax and slavery," proclaimed a 2001 article in the right-wing magazine *Human Events*.[16]

Complaints about taxes are bad news for the IRS. Scholars who have examined "the psychology of taxation" say that a tax system is in big trouble if it lacks legitimacy. Distrust in a society also spells trouble for tax collectors. Research across nations has found that tax evasion is higher in societies with lower levels of trust between people. People must *want* to pay taxes at some level, believing not only that their tax bill is fair, but that their destiny is bound up with that of their fellow citizens. A great many Americans, however, don't feel this kind of social solidarity.[17]

Of course, tax evasion is not just about legitimacy. Such beefs may be a prime motive for some tax cheats and not a motive at all for others. Many people cheat simply because "everybody" cheats and few people get caught. (A 2003 poll found that 95 percent of Americans believe that "other people" cheat on their taxes.) "My rationalization is that everybody does it," says a woman from the Midwest. Like Jennifer Bennett, this woman is a straight arrow. While it's hard for her to imagine shoplifting a pack of gum, she doesn't think twice about committing a felony every April. "It's the game you play at tax time, and you're kind of a sucker if you actually report accurately and honestly."

The fear that following the rules makes you a chump can lead to what experts on academic dishonesty call the "cheating effect." People otherwise not prone to cheating come to do so because they don't want to needlessly pay more or put themselves at a disadvantage. Arguments that "everybody does it" serve as a key rationalization for the tax cheating of ordinary middle-class Americans, as well as other kinds of cheating. The pervasiveness of this rationalization shows how easily cheating can create a downward spiral: The more cheating there is, the more it becomes a routine part of life. The more it's normalized, the less it becomes a conscious choice driven by any meaningful motive at all.

What is tricky in these situations, and often impossible, is tracing this cycle back to its origins and making judgments about underlying causes. Why did "everybody" start doing "it" in the first place? Why didn't somebody stop the cheating before it became commonplace? Why is so much cheating tolerated, even though the authorities recognize its prevalence?

Data is not available to fully answer these questions when it comes to tax evasion. For example, nobody has ever conducted regular surveys over a period of years that asked large numbers of tax evaders why they cut corners. Here, as with many other forms of cheating, we can only speculate about what is going on by synthesizing our knowledge of larger trends and attitudes with anecdotal accounts of cheating.

My own informed guess is that economic anxiety and perceptions of unfairness serve as prime rationalizations for many middle-class tax cheats, maybe even for most. And it can be said far less speculatively that the attack on government during the past two decades is what largely accounts for the IRS's weak enforcement of the tax laws. "Everybody" cheats on their taxes because they can get away with it. And why is that? Because government enforcers, deliberately starved and hobbled, have had little choice but to allow so many people to do so.

PERVASIVE EMPLOYEE theft by ordinary Americans is another problem that is tricky to unravel. But here, too, you can see at work the toxic cynicism that fuels cheating in a society rife with unfairness.

Employee theft is actually the biggest area of cheating by the middle class—and it's been rising fast in recent years. This includes everything from taking home pens from work to padding one's expense account to stealing outright large sums of money through insider financial schemes. The Fraud, Forensic & Investigative Services Group at Ernst & Young estimated the total cost of employee theft and fraud at a staggering $600 billion in 2002—or $4,500 for each U.S. employee and 6 percent of the national GDP.[18]

The biggest losses to an organization occur through high-level white-collar thefts, but the sheer scope of wrongdoing by lower-level employees also adds up to a great deal of money. Taking into account only those employees who got caught, one in every 22.4 retail employees stole from his or her employer in 2000. A study sponsored by the National Food Service Security Council found that restaurant employees stole an average of $204 each in cash or merchandise in 2000—more than twice as much as the average take in 1998. Business losses to check fraud by employees, including forgeries and mailroom theft, doubled between 1994 and 1998, with firms surveyed in a KPMG study reporting an average of $624,000 in losses. The same survey revealed that misuse of ATM cards had resulted in an average loss of $300,000 per firm—once again doubling 1994 losses in the category.[19]

All these trends, it should be noted, occurred at the same time that shoplifting and other street-level property crimes were decreasing. On the other hand, this was a period of rising disparities in pay, growing consumerist and cost-of-living pressures, the spread of harsh new management ideas like "forced ranking," and a decline in the security and fringe benefits offered by many jobs.

Credit cards are a favorite vehicle for employee abuses. The company expense account presents a unique temptation to live a financial life other than one's own, to close the gap between material expectations and real-life opportunity. It is easy for those with company credit cards to get carried away; and it's also easy for those who don't have company cards to get hold of the card numbers and engage in unauthorized spending. According to the KPMG survey of employee theft and fraud, abuse of company credit cards increased by a factor of three between 1994 and 1998, resulting in an average loss of $1.1 million per firm. Expense account fraud, meanwhile, averaged $141,000 per firm in 1998—seven times as much as KPMG had identified in its 1994 survey.[20]

There are no surveys that reveal the motives of thieving employees, and it is risky to generalize about people's rationalizations for their behavior. But perceptions of fairness do seem to play a big role here. "If someone feels inadequately compensated or poorly treated, they might look for other compensation," comments Joseph Wells, who is founder and chairman of the Association of Certified Fraud Examiners, and a former FBI agent who worked the white-collar crime beat. Wells may know more about patterns of employee theft than anyone in America. "Companies who treat their employees badly generally have a bigger fraud problem," he says.

Another expert, a lawyer who works with Wells, puts the situation this way: "You see the guy down the street who cashed in his stock options and is now worth a million dollars, and you're still making $30,000. . . . It makes it easier to rationalize fraudulent behavior."[21] Recall the bookkeeper and the $100 a month. Embezzlement or chiseling offers employees a chance to give themselves the compensation they believe they would get if the world were fair.

Other thieving employees may fancy themselves as modern-day Robin Hoods. A New Jersey man who worked for a large financial services company described to me how he stole the equivalent of tens of thousands of dollars from his employer—all for a good cause, he insisted. He worked full-time for the company in a computer job that actually required minimal attention. "The job was a joke that took about half a brain cell," he says. So every week for several years straight, he spent much of his time in the office working on his true passion—his religious volunteer activities. And he went further than just theft of services. He also used company office supplies to photocopy and send out huge mailings for several conferences focused on troubled youths that he organized on a shoestring budget. "I wouldn't have been able to put together the conference without company stuff," he says. "I did a photocopying job that amounted to 18,000 sheets of paper." Late one night, alone at the photocopier, he ran into a company lawyer who looked at him askance but ultimately didn't raise any questions. At one point, he stacked boxes of photocopied materials in his office, in plain sight of his supervisor, but she never asked any questions.

"Nobody cared," the man said. "A lot of the people in my department cut corners in one way or the other and had the attitude that it was okay to live off the fat of the company."

While many companies got leaner and meaner in the '90s, this particular company tolerated plenty of fat as it raked in record profits during the bull market of the '90s. Like many financial services companies, this one was engaged in some shady practices during the boom, which were no secret to the employees who worked there. "I certainly didn't care about ripping them off," says the man. "You had the CEO making $50 million a year, and he and all the others are lying to investors. And there I am making $50,000 a year and trying to do something worthwhile in the world."

As with tax evasion, it's hard to say how much employee theft is rationalized by ideas of fairness and how much is just unthinking criminality. Polls paint a complicated picture of why Americans may or may not be prone to steal at work. While the majority of Americans say they are happy with their jobs, other evidence points to a more jaded and even poisonous mood in the workplace. Only two thirds of workers say that their employer is loyal to them even as over four fifths say they are loyal to their employer—quite a loyalty gap. And, according to a 2002 survey of 13,000 workers, just 39 percent of employees trust senior managers. Another survey found that less than half of employees felt that "their senior leaders are people of high integrity." Polls also find that large percentages of employees say they've witnessed something illegal or unethical happening at work over the past year. At the same time, only half of workers say they get a sense of identity from their job, a third say they'd be happier in another job, and nearly half say they are stressed out a great deal of the time at work. Many workers also gripe about how much money they make (not enough), their health insurance benefits (or lack thereof), and their chances for promotion (poor). While job satisfaction is relatively high, it has been declining over the past decade, and loyalty among most workers is paper thin: One 2002 survey found that only a quarter of workers plan to stay at their jobs for at least two years and that half wouldn't recommend their company to someone else. Barely a majority of workers in this survey said their employer treated them fairly and, on balance, the survey classified only 24 percent of workers as "truly loyal." Millions of Americans, in short, don't arrive at work every day with what might be called a positive attitude, and this might explain why employers are being ripped off left and right.[22]

Whatever the motive for employee theft, economic booms have historically been a good time to get away with bad behavior at the office. "At any given time there exists an inventory of undiscovered embezzlement," wrote John Kenneth Galbraith in *The Great Crash of 1929*. But this inventory tends to be higher during booms. "In good times people are relaxed, trusting and money is plentiful. But even though money is plentiful, there are always many people who need more.... Under these circumstances the rate of embezzlement grows, the rate of discovery falls off, and bezzle increases rapidly. In a depression, all this is reversed. Money is watched with a narrow, suspicious eye."[23]

The 1990s saw a subtle variation on this theme: things were indeed flush, but employees were also under mounting financial and work pressures during these "good times." In other words, there was plenty of money sloshing around for the taking, and people had plenty of motives to grab it.

KEEP FOLLOWING the logic about middle-class cheating and consider music piracy over the Internet. Different rationalizations blend in even murkier ways here than elsewhere—and yet beefs about fairness are never too far from center stage.

Napster debuted in 1999, the brainchild of a college dropout named Shawn Fanning. The software program allowed users to search hard drives of other PC users for the music they wanted and then download that music into their own computer. Napster's use spread like wildfire, especially on college campuses. It had 65 million registered users by 2001. Other file-swapping programs quickly appeared, including Kazaa, Morpheus, and Grokster. Napster was shut down by court order long ago, but other programs continue to thrive. Kazaa has over 100 million registered users. There are now several million people engaged in music file swapping at any given moment of any day, and it's estimated that 2.6 billion music files are downloaded every month, mostly in a way that is a federal felony. Surveys show that nearly 50 percent of teenagers and a quarter of all Americans have downloaded music in the past month. Forty-two percent of admitted music pirates report having copied a CD rather than buying it.[24]

Fanning defended his invention by arguing that Napster actually was good for the industry in that it allowed consumers to sample different kinds of music "in a way that gives them control over how they experience it," thus creating more demands for CDs. This argument may have had plausibility early on, and initial survey research of music file swappers seemed to confirm the claim. But there is now unambiguous evidence that Internet music piracy is hurting the recording industry. Sales of CDs have dropped over the past several years and research links much of this drop to music piracy. However, at least one kind of CD is selling more briskly than ever: 1.7 billion blank CDs were sold in 2002, an increase of 40 percent over 2001. Internet music piracy, in short, represents intellectual property theft of major dimensions.[25]

Big companies like Sony are taking a hit and so are musicians. In some cases where the hot song on a band's new CD has already been widely pirated over the Web, the consequences for sales of the CD can be serious. Such incidents are unnerving. In the unforgiving music business, a single CD that flops can spell permanent failure for a band. While some recording artists have embraced music swapping as a path to larger audiences, many artists agreed with the leaders of Metallica when they compared music swappers to "common looters loading up shopping carts because 'everybody else is doing it.'" Metallica's drummer, Lars Ulrich, added this about music swapping: "It's not the Metallicas... that take the hardest hit, it's the 10 developing bands each label has on its roster every month. That gets trimmed to three.... Young artists won't have a chance."[26]

There is nothing extraordinary about people who pirate music over the Internet. Although the computer world has always had a criminal hacker element, music piracy doesn't rise to this level. Music piracy isn't like hacking into the Strategic Air Command or Citibank. Felony or not, it's a mundane and routine activity for millions of Internet users.

And how do these users justify their thieving behavior? "I don't justify it," said one music pirate with a shrug. "I haven't really thought about it." This nonchalance is not atypical. Surveys show that less than 10 percent of music swappers think that piracy is wrong—even though 20 percent believe it hurts artists and not just the record companies.[27]

Those who do bother justifying their actions offer several rationalizations. Some music pirates echo Fanning's original defense of Napster, saying that the Internet has opened their eyes to new music. "It's a chance to discover new bands and new songs," says another veteran music pirate. "And once you know you like something, you feel better about buying the whole CD." This man draws the line at burning CDs. "I'm against people downloading a whole CD.... I feel better if I go out and buy a CD."

More legalistic rationales—used by Napster's lawyers when record companies went on the warpath—focus on the rights of individuals to make copies of music or videos for their own personal, noncommercial use as affirmed by the Audio Home Recording Act and a 1984 Supreme Court ruling. One is allowed to make a digital copy of a CD one owns and store it on one's home computer, and also to give the file to a friend—just as it is legal to mix a cassette tape and give it to a friend. Music file-sharing programs simply facilitate this long-established behavior on a much greater scale.

But scale is exactly the catch here. It's one thing for people to occasionally pick a flower in a public park; that shouldn't be criminal. It's another thing if numerous people go into the park every day to assemble bouquets. Because "everybody" can now get free music, occasional sharing has snowballed into mass stealing. Recent court opinions have endorsed this distinction in a blow for music swappers.

This is where the second defense of music piracy comes in: that such stealing is okay.

"It's a rebellion against the high price of CDs," says a music pirate who works as a day trader on Long Island. "Artists aren't making the money, just the CD makers." Mick Brady of *E-Commerce Times* voiced a commonly held view about record company losses when he said it "serves them right after abusing artists all these years" and also mistreating consumers. Music swappers, he said, "rose against oppression."[28] Other commentators have compared music piracy to the Boston Tea Party and the Civil Rights movement—as just disobedience against an unjust regime. College students in particular see themselves as victims of the music industry's greed. The price of a CD has risen to as much as $18 or more at the same time that the music industry's huge young consumer base is struggling with rising college tuition costs and escalating student debt burdens.

Music swappers fighting "the power" also stress that they are liberating high-quality music from the clutches of corporate America. "It's harder and harder for small bands to break out these days given the dominance that conglomerates have over what gets played on the radio or on MTV," says a music pirate. Music swapping creates an audience for bands that would otherwise languish in obscurity. Never mind that such swapping can also doom any commercial prospects the bands might have.

The music industry is fighting back against piracy. Ominous warnings are popping up on the screens of music swappers that spell out the legal consequences of hitting the download button. Subpoenas are being dispatched to egregious offenders nationwide. Universities are also beginning to crack down on the problem, trying to stay one step ahead of the recording industry, which is increasingly targeting student music swappers. One student at Michigan Tech was sued by the Recording Industry of America for $98 billion for indexing 650,000 songs on an Internet site. He settled out of court for $15,000.[29]

The recording industry is fighting an uphill battle, especially with more homes getting broadband Internet connections. Its only hope is that music swapping can be channeled into legal venues, such as Apple's new iTunes initiative for sharing files. Like other forms of low-level cheating, music piracy long ago passed into the "everybody does it" zone and is now utterly normal. "Ripping off the music industry is just part of the culture," says the day trader, whose nighttime activities continue to include music piracy.

Cable television theft and auto insurance fraud get a lot less attention than music piracy, but these problems are much more costly. Industry estimates peg the annual losses from cable television theft at more than $6 billion. Cable piracy can diminish the picture quality for cable subscribers by weakening the signal moving through a cable network—resulting in unhappy customers and more service calls. It also may result in higher rates for subscribers than would otherwise be the case, as the unrealized revenues from potential customers who cheat are made up through charges to actual customers.

Cable theft hardly tops the list of America's most pressing national problems—or even makes that list. But it is not a victimless crime, and its pervasiveness offers yet more unsettling news about the state of American ethics.

Wherever there is cable, it seems, there is also theft of cable services. It is estimated that 100,000 households steal cable service in Chicago alone, or roughly a tenth of households. Cable theft is so much a part of the fabric of American life that it has been the subject of plotlines in television shows like *Seinfeld, The Simpsons,* and *Mad About You.*[30]

The cable pirate generally sees himself as a law-abiding citizen. "I think for a lot of people, this is the only illegal act they ever do," commented a city official in Chicago who launched a special initiative to go after cable pirates. In one case in New Bedford, Massachusetts, it was revealed that the police department was receiving illegal cable television programming. The cops were getting not just the New Sports Network channel without paying but also Spice, a premium channel that carries adult content. Nobody in the station house thought much of this scam until AT&T investigators showed up.[31]

The justifications for cable theft track closely with those for music swapping. The cable television industry has even more enemies than the music industry. Cable providers often have a local monopoly, and they are loathed for any number of reasons: because they overcharge, because they don't provide good service, and because they are rude to customers. Cable horror stories are a staple of American chitchat, and have even inspired a Jim Carrey movie, *Cable Guy,* about a whacked-out cable serviceman who causes chaos in an ordinary American household. Much of the hostility toward cable companies can be traced to the corporate takeover of this industry. As one industry analyst commented: "While some cable systems—particularly those that were once small, independent operators with deep roots in their service communities—made customer service a religion, many lost the true faith as they were gobbled up by increasingly large competitors."[32]

Exacting revenge against tyrannical cable companies strikes many Americans as a just cause if ever there was one. Those engaged in cable theft regularly cite the misdeeds of cable companies to justify their behavior. A favorite excuse uses the logic of expropriation—that the cable companies are making so much money by gouging consumers that they should be forced to pay some back. This same logic can be found among employees who steal from their company and among music pirates.

And then there are those who just don't give cable theft much thought. "I didn't go out looking," said a Chicago man busted for cable theft after he paid a cut-rate price to get an illegal hookup. "I was talking to some guys and I just said: 'Yeah, why not.'"[33]

More Americans are also saying "why not" when it comes to ripping off their car insurance company. And here, too, when they do think about their cheating, they justify it with righteous anger of a hated system.

Auto insurance is like a dull backache in many people's lives. Insurance rates have climbed to unbearable levels in various places and seem to always be going up. Americans spend $100 billion annually on auto insurance, and it's not uncommon for a household with new two cars to be paying more than $3,000 a year. Exorbitant rates rise even higher for those with a tarnished driving record, and insurers are inevitably uninterested in the fine points of why a driver got ticketed. Auto insurers are also notorious for bumping up the rates of people who've been in accidents—regardless of where the blame lies. This practice makes many people hesitant to use their overpriced insurance when they actually need it. Nearly every day, gripes like these are vented on www.screwedbyinsurance.com, a consumer complaint Web site that gets plenty of traffic.

Auto insurance is required by the government and by banks that make loans on cars, but many states don't strongly regulate an auto insurance industry whose many lobbyists buzz around state capitals like flies. This leaves captive consumers to get gouged by a small group of leading insurers who often enjoy healthy profits at the expense of millions of policyholders. In California, for example, a study by a consumer group found that profiteering by the insurance companies led state drivers to pay $5.2 billion more in auto insurance than they should have between 1995 and 1997.[34]

High insurance premiums take a bite out of everyone's finances, but they fall particularly hard on middle- and lower-income drivers—serving, in a car-centric society, as a regressive *de facto* tax. While rising health-care costs have gotten a lot more play in the media, the rising burden of auto insurance is another part of the squeeze on working families.

Auto insurance fraud of an explicitly criminal nature is a major problem in the United States, and includes staged accidents, bogus claims, and the like. But low-level cheating around auto insurance by otherwise law-abiding citizens is also pervasive and there is evidence that it has grown in the past decade. This cheating takes various forms. For starters, many people lie about who is driving the insured car. If a wife has a good driving record and the husband has a bad one, both family cars may get registered in the wife's name. Or maybe the presence of a teenage driver in the house won't be disclosed—a piece of information that can change the price of a policy by hundreds of dollars.

Then there is lying about the location of the insured car. Many people mislead insurers with the help of friends and relatives who live in cheaper insurance zones. Those with two homes have an easier time. An L.A. Mercedes is registered in Malibu; a Boston Lexus is registered in the Berkshires; a Land Cruiser garaged in Manhattan is registered in East Hampton, and so on. In one Catskills county ninety miles north of New York a district attorney estimated that out of the 74,000 vehicles registered in the county, 5,000 belonged to people from outside the county, mostly New Yorkers.[35]

More advanced auto insurance cheating involves lying about claims. A car owner who skimped on collision insurance or went for a higher deductible will finally cough up the money for better coverage—after the damage has occurred that he wants to get fixed. I did this myself once with a cracked windshield, at the suggestion of my friendly Allstate representative. Or, on a bigger scale, people make false claims about injuries in auto accidents. This kind of fraud alone may cost insurers over $6 billion a year. A harmless fender bender, it seems, is yet another way to make a quick buck.[36]

In a 1991 public opinion survey about attitudes toward auto insurers, large numbers of respondents said it was okay to lie and cheat about car insurance. A third said it was okay to lie about the number of miles they drive each year in order to get a lower premium. A quarter said it was okay to lie about their address. Others endorsed such cheating as lying about one's driving record, lying about the value of a stolen car, and lying about the scope of medical problems caused by an accident.

"A good many of these people probably don't consider these things fraud," said Donald Segraves, executive director of Insurance Research Council, an industry group. "That's part of the problem. To them, it's like fudging a bit on the speed limits. They think it's OK to fudge a little, and they do not regard themselves as being criminal or perpetrating a fraud."[37]

Auto insurance, in other words, is just one more area where the cheating culture flourishes.

Is AMERICA TURNING into a corrupt nation akin to Brazil? Is the rising tide of low-level cheating around taxes, workplace theft, car insurance, and cable television eroding the fabric of national life?

It depends on how you look at the problem. All this low-level cheating doesn't add up to a pressing crisis—though it certainly makes us all pay more to subsidize those who cheat. But what's most troubling is what it says about the cynicism of the American public and the ways that cynicism acts as a corrosive agent on people's integrity.

Who's to blame for all this cynicism? Mainly those at the top of our society.

America's most powerful and wealthy individuals have always shaped our culture in innumerable ways. This is truer now than ever, with the media devoting more attention to the doings and opinions of VIPs. At various earlier times, elites took the responsibility that comes with influence and visibility seriously. They embraced notions of noblesse oblige, a commitment to public service, and informal edicts against conspicuous consumption. None of this is to say that yesterday's elites were angels; they weren't. But life at the top has grown much uglier since the '70s. The past quarter of a century has seen more scandals among the rich and powerful than any period since the Gilded Age. Again and again, celebrities that millions of Americans worship turn out to have dark sides. For instance, consider the letdown of Martha Stewart's fans in the heartland as they read biographies of her that are jam-packed with shocking stories about horrible behavior—the long trail of betrayals that fueled her rise in business, the vicious feuds with neighbors in Westport and other places where she has homes, and now a federal indictment for lying to prosecutors. Or think of the example that slugger Mark McGwire set for young athletes when he admitted using steroids the year he broke Roger Maris's home-run record. Or think of how amateur investors—ordinary people trying to grow their nest eggs—must have felt when it was revealed that Henry Blodget, a CNBC guru that everyone loved, was pushing stocks he privately bad-mouthed. Think of how these investors feel now, with their nest eggs hollowed out and Blodget sitting on a huge personal fortune.

The actions of the Winning Class send a message to the Anxious Class. The message isn't just that the world is unfair and the rich can get away with murder; it's that people who cut corners get ahead. Most of us don't get a chance to cut corners in any big way. We don't get the chance to earn $12 million a year by lying to millions of people on TV, or to make a killing by cashing in on insider trading tips. Yet we get plenty of small chances to make an extra buck. And now it seems that more than ever, we seize those chances. Should this really come as any surprise?

America is a long way from becoming Brazil, and many areas of American life are less corrupt than they were fifty years ago. Government operates more cleanly than it ever has, leaving aside all the legal bribery of the campaign-finance system. Labor unions are also less corrupt, while the influence of organized crime has declined notably in many sectors. And believe it or not, stronger ethical codes have been enacted in nearly all of society's major institutions over the past quarter century—especially in the business world. Another major victory for fairness and equal justice has come from the triumphs of feminism and multiculturalism since the '60s, which have slowly weakened the exclusionary bonds of old-boy networks in numerous sectors of society.

Still, amid all these important gains, "the Brazilianization" of American society marches forward and could grow worse.[38] If vast income gaps are left unaddressed, as they are in Brazil; if so many Americans continue to feel economically insecure, as so many Brazilians do; and if a two-tier system of justice continues to prevail—one for the rich and one for everyone else—as it does in Brazil, more and more people will question the basic legitimacy of the social contract governing our society, as many long have in Brazil. More of us will make up on our own ethics and rules. And more forms of cheating will become so commonplace that people won't even think to justify their behavior.

And one day, perhaps not so far in the future, Americans may wake up and realize that they are living in a place quite similar to Brazil—but without the great beaches.

CHAPTER SEVEN

Cheating from the Starting Line

Just north of Manhattan, in what is technically the Bronx, lies the tony neighborhood of Riverdale. The area boasts Georgian mansions and ivy-covered apartment buildings with majestic views of the Hudson. Riverdale is also home to the eighteen-acre campus of the Horace Mann School, one of the most exclusive private day schools in the United States. Horace Mann might seem to be a place where young people would take for granted a prosperous future. The 1,000 students, ranging from two-year-olds in nursery school to high school seniors, come from some of the wealthiest families in America. If anyone has a shot at success it is these kids—with their trust funds, their parents' connections, their six-cylinder birthday presents, and their bedrooms jammed with the latest high-tech gadgets.

A Horace Mann education by itself would seem to be a ticket to the Winning Class. For a tuition of around $20,000 a year, the school offers a cornucopia of educational opportunities designed to prepare students to compete in the new economy. There is a sophisticated computer curriculum, instruction in seven languages, an award-winning weekly newspaper (the school has produced twelve Pulitzer Prize–winning alums), a visual and performing arts program with ties to leading New York cultural institutions, and nearly fifty student clubs. The school's grounds include extensive playing fields that are immaculately maintained, an elaborate gymnasium complex, and seven tennis courts. Students can compete in twenty-three different sports. Horace Mann also has a crack team of college counselors, and the 110-year-old school long ago earned the nickname "Harvard Man." In past years, after the annual senior prom at the ritzy Pierre Hotel, as many as half of all graduates have gone on to Ivy League colleges.

Horace Mann is home to some very accomplished cheaters. For example, there's the star student who bought a term paper off the Internet to hand in for one of her classes. This in itself is not unusual. Downloading papers from the Internet, or plagiarizing parts of papers from the Web, is common at Horace Mann and many other high schools. The problem is that it's so common that sometimes two students will find the same paper on the Internet and hand it in at the same time, which is what happened to this particular girl. In talking about the assignments with her classmates, she found out that someone else had handed in the same paper. She was horrified. She had big plans to apply to top Ivy League schools. A scandal could throw a wrench into things. She immediately rushed to the teacher and begged to get the paper back, saying she needed to fix it. She never got caught. "She goes to Harvard now," says a recent Horace Mann grad who related this story. "These people learn how to work the system."

One student won notoriety for hacking into the school's computer system to change his grades. In another case, a group of students developed an elaborate system of signaling to swap answers during tests. Horace Mann is rife with rumors of other sophisticated cheating scams, like special pens and pencils that are modified to hold cheat sheets. All of these tricks are designed for one purpose: to get into a top college.

"What Horace Mann programs you to think is that your purpose in life is getting into an Ivy League school," says another former student, who graduated from the school a few years back. "Success is purely numeric." Students are obsessed with their grades and are driven relentlessly to succeed. In many cases this drive is deeply ingrained in their family lives and immediate community. "There is a story that when I was seven I told my dad I was going to go to Princeton and that my two best friends would be my roommates," comments a third recent Horace Mann grad, who began attending the school at the age of three.

"I had one friend in particular who never did his own homework," this grad goes on to say. "He copied my homework every day. He may have bothered some people because he was a smart kid who didn't do anything. He was really good at cheating; he managed to cheat on the SATs. He was taking it untimed because of a learning disability. He would bring a pocket dictionary into the bathroom. He goes to Harvard now because his parents gave a lot of money. I don't see this as something that's going to change. The students at Harvard benefit because it has the biggest endowment in the country."

Some teachers at Horace Mann fan the competitiveness over grades and college admissions, at times openly stating the grades of students in class and publicly humiliating students who do poorly. Other faculty at Horace Mann acknowledge that the intense focus on grades and Ivy League schools goes too far, but say the school isn't to blame for these excesses. "Parents and kids fall back on brand names," the school's lead college counselor once explained, "believing it is their security in an increasingly insecure economy."[1] The obsession with guaranteeing one's advancement in the world, along with the exclusive focus on ends, can easily justify dishonest means—not only at Horace Mann but at other schools, too. "You can forget what you're doing and think this is a very small thing and it will get me an A," says the student who was imagining his life at Princeton at age seven. "You feel that it's worth it, you're not stealing a test and passing it out, but if you copy a number off of a person next to you, you figure, it's not going to change how much I know at this point, but it will affect my grade, so why not?"

Thirteen miles south of Horace Mann, not far from a vast asphalt pit that used to be the World Trade Center, is another high school and another pressure cooker. But this one is filled with a very different kind of student. Stuyvesant High School was founded in 1904 as a "manual training school for boys." Over the years it evolved into something else entirely, and is now New York City's premier public high school—some even say the best public high school in the country. Stuyvesant has a new multimillion-dollar building overlooking the Hudson River that is almost as well equipped as Horace Mann, with laboratories and ubiquitous computer terminals, and facilities for its thirty athletic teams. Stuyvesant is known for its award-winning science students and killer debate team. Every year many of its students go on to the Ivy League.

Yet the school is anything but a bastion of privilege. Stuyvesant is a meritocracy of the rawest sort. It admits some 800 freshman students every year by a competitive examination that is open to any eighth grader in the city. Along with the Bronx High School of Science and just one or two other decent public high schools, Stuyvesant beckons as a path to economic security—in many cases the only possible path—for kids from middle- and lower-income New York families. Its building on the Hudson is a place of exalted dreams and of pent-up aspirations often traceable just a few decades back to peasant cultures abroad. But

Stuyvesant is also a place of profound desperation and extreme pressures.

Two decades ago, Stuyvesant's students were mainly Jewish or Eastern European. Back then, the school was located on East 15th Street, and it was filled with kids from neighborhoods in the East Village and Upper West Side. These days, New York's Jews and Greeks and Hungarians and Romanians have largely graduated to secure spots in the middle or upper middle class; most send their children to private schools. Stuyvesant today is filled with a new generation of strivers who are clawing their way up from the lower rungs of the economic ladder—Indian kids from Queens, Chinese kids from Chinatown, Koreans from all boroughs, and others: West Indians, Pakistanis, Vietnamese, Bengalis. Half of Stuyvesant's students are Asian Americans; many are from first- or second-generation immigrant families. It's not uncommon for Stuyvesant students to juggle part-time jobs on top of their schoolwork to help support their families.

The demanding nature of a Stuyvesant education is notorious. "The pressure at Stuyvesant was immeasurably greater than the pressure at Harvard," says Ben Shultz, who attended both schools. Students at Stuyvesant are expected to excel at all subjects—math, science, English, history—and there is no leeway to slide through the system by playing to one's strengths. Many students stay up late at night dealing with their crushing homework load and then walk around all day in a sleep-deprived daze. They take caffeine pills and amphetamines to keep themselves going. The amphetamine problem is serious enough that the school has sponsored seminars and lectures to raise student awareness about the drug's health risks.

The hardcore atmosphere at Stuyvesant intensifies at exam time, turning even more deadly serious. Quite apart from the pressures on students from themselves and teachers, many are burdened by parents who push them relentlessly—parents who discuss every grade on every exam, who know their kid's GPA as well as the kid does, who see academic success at Stuyvesant as the culmination of generations of struggle.

How do Stuyvesant students deal with these pressures? Often by cheating. "The whole culture was about cheating," says a former Stuyvesant student who returned after graduating to work as a guidance counselor at the school. "Everybody cheated." The atmosphere encourages it in multiple ways. The academic emphasis at Stuyvesant is on preparing students for the many tests that will determine their future: state regents tests, advanced placement tests, SAT tests, and the endless quizzes and exams that are the basis of grade point averages that students track obsessively.

"There was little focus on intellectual curiosity and only a focus on grades," says the former student. "Everybody was trying to get into Harvard and Yale. It was made very clear that that was the only way to succeed in life, and that you'd be a total failure if you didn't get into the right place." The emphasis on college admissions starts very early on at Stuyvesant. "In the beginning of freshman year, they sit you down and have a talk about starting to prepare for college applications," says Martina Meijer, a recent graduate. "They talk about keeping your grades up, and doing as many activities as you can. Everyone knows their average down to the hundredth point, which is sad. As a sophomore, I distinctly remember calculating how each test and quiz would affect my overall average."

The cheating culture at Stuyvesant, as at Horace Mann, has created folklore that is passed down among students and which grads remember many years later. Former students talk, for example, about the law of rising test scores. "There was a well-known pattern that people who took exams later in the day would do better on average," says Jesse Shapiro, who graduated from Stuyvesant in 1997. It was considered the civic duty of morning students to fully brief afternoon students on the content—and answers—of tests in a given class. Students who were savvy, and unscrupulous, were wise to take courses in their weaker subjects in afternoon slots.

Another pattern suggested that supersmart students were the suns around which cheating planets revolved. Shapiro recalls a science teacher illustrating this pattern by drawing an isobar graph, or contour map, on which he plotted the grades of an entire class to show how good grades radiated from certain points in the room.

New technological gadgets have introduced fresh legends into Stuyvesant's cheating folklore. Handheld text-messaging devices allow students to shoot test answers across the school, or even across a class, in seconds. Calculators, which students can often bring into math tests, are programmed in advance to contain answers. E-mail allows students to more easily share homework answers with each other during evening hours, while the Internet enables students to engage in the common practice of downloading papers and passing them off as their own. "Few original papers get written these days," commented one former Stuyvesant student.

Not all Stuyvesant students cheat, of course, and there is no solid data on which students are most inclined to cheat. But people have theories on this matter. Douglas Goetsch, a teacher at Stuyvesant, wrote an article for the school newspaper on the problem of cheating. He concluded from his own experience that nearly all the students who are cheating are those with an "excessively demanding parent."[2]

The knowledge that some students are cheating creates angst on the part of other students and may fuel their own cheating—what researchers call "the cheating effect." Students at Stuyvesant perceive college admissions as a zero-sum game in which another student's gain is your loss. "You're simply competing with your classmates for a spot at a school," says Meijer, who is now at Amherst. "Harvard only takes a certain number of students from Stuyvesant each year."

HORACE MANN AND Stuyvesant High Schools are unusually competitive places, but more and more students are under the kinds of pressures found at these schools. Parents and students understand that the stakes of education have shot up in recent years. A growing obsession with college admission has been paralleled by increased cheating among high school students across the United States. According to large-scale national surveys by the Josephson Institute of Ethics, the number of students admitting that they cheated on an exam at least once in the previous year jumped from 61 percent in 1992 to 74 percent in 2002. "The evidence is that a willingness to cheat has become the norm and that parents, teachers, coaches and even religious educators have not been able to stem the tide," commented Michael Josephson, the institute's president.

Surveys by the institute have also found that more students say that they "sometimes lie to save money" than was the case in 1992, and that more are willing to lie to get a job. In addition, these surveys confirm the importance of financial success among today's young people. High school students rank "getting a high-paying job" above "being ethical and honorable," above following current events or participating in politics, and—most surprising—above being attractive or popular. Even as they place extreme importance on financial success, high school students also increasingly believe that "a person has to lie or cheat sometimes in order to succeed." Forty-three percent of the 12,000 high school students surveyed in 2002 agreed with this statement, up from 34 percent in 2000.

Nearly 40 percent also admitted that they were willing to lie or cheat to get into college. Do these students feel bad about all the corners they are willing to cut in life? Not at all. Three quarters of high school students said that they were more likely to do the right thing than most people they knew and 91 percent agreed that "I am satisfied with my own ethics and character."[3]

Another large survey of high school students, the annual opinion study *Who's Who Among American High School Students,* offers some additional insights into the problem of cheating among high schoolers. The *Who's Who* surveys have been conducted annually for thirty years and they have focused only on top high school students with good grades and college aspirations. The study first began asking about academic cheating in 1983. Since then, the number of students who admit to cheating at some time during their academic career has increased from 70 percent in 1983 to 80 percent in 2000.[4] The 2000 *Who's Who Among American High School Students* reported a record number of cheaters with A averages, with 80 percent of these students admitting to some form of academic dishonesty. Not only was the percentage of self-identified cheaters the largest ever recorded in the *Who's Who* survey, but most students indicated that cheating was "no big deal."[5]

Interestingly, America's top high school students aren't breaking rules in other areas. A review of findings from *Who's Who* over a twenty-five-year period concluded that teens "have over the years become more responsible and more mature about taking charge of their lives: fewer teens drink, smoke, or use marijuana, and more of sexually active teens use contraceptives these days."[6] This divergence underscores the lopsided nature of America's moral conversation over the past two decades. A nearly exclusive focus on drugs, sex, and crime has helped to change behavior among young people in these areas. But there has been little attention paid to problems like greed, materialism, and excessive competition. Young people seem to be hearing "just say no" about some temptations—and "do whatever it takes" about others.

THE ADMISSIONS OFFICE at Harvard College is a button-down, secretive place. It is accustomed to withstanding entreaties on behalf of applicants from some of the most powerful people in the world—from foreign leaders to U.S. senators to top CEOs. Discretion is a part of life for the admissions team, and it is not known for its rash pronouncements. Yet in 2000, the dean of admissions took the highly unusual step of publicly castigating America's parents for how they primed their young to succeed. In an essay co-authored with the director of admissions and another university official, Dean William Fitzsimmons put the college admissions craze in the context of broader trends: "Stories about the latest twenty-something '.com' multimillionaires, the astronomical salaries for athletes and pop-music stars, and the often staggering compensation packages for CEOs only stimulate the frenzied search for the brass ring.... More than ever, students (and their parents) seek to emulate those who win the 'top prizes' and the accompanying disproportionate rewards."[7]

The essay by Fitzsimmons and the other Harvard officials went on to offer a scathing analysis of elite educational competition—from cradle to college. Brutal competition begins even before kindergarten, the authors noted, when parents start jockeying to get their child into the right preschool. "The competition for admission to some of the Pre-K, Kindergarten, and grammar schools," they wrote, "can be...statistically more difficult (with lower admission rates) than Harvard."[8] Consultants are paid huge sums of money to coach and tutor preschoolers, including rigorous prepping to help the children impress interviewers at prestigious preschools with their ability to make eye contact and play nicely with others. At the same time, parents pull every conceivable string to get their child into the right school. (The essay by the Harvard officials was published well before the revelations about a darkly comic episode of preschool corruption involving Jack Grubman, the Wall Street telecom analyst who allegedly upgraded AT&T's stock in an effort to get Citigroup chairman Sanford Weill to help Grubman's twins gain admittance to the 92nd Street Y preschool, one of the most elite preschools in New York City.)

The practices of parents become even more corrupt later, when college appears on the horizon. Not just any college will do for many parents. It must have a name lustrous enough to inoculate its grads against the insecurities of the new economy and serve as a stepping-stone to the Winning Class. As the Harvard admissions team writes, professional college counselors "appear on the scene early, sometimes in middle school, to begin to structure students' academic and extracurricular profiles for entrance to the 'right' college.... From a cynical perspective, such advice steers students toward travel abroad, community service, or other activities solely to enhance college essays or interviews."[9] One service in New York, Ivy Wise, offers a "platinum package" of college counseling services that costs nearly $30,000 and consists of twenty-four counseling sessions for a high school student beginning in eleventh grade. Similar services are cropping up in wealthy areas across the U.S. Whereas only 1 percent of college freshman admitted to consulting a private admissions counselor in 1990, 10 percent of today's students say they benefited from hired help.[10]

While many admissions counselors are ethical in the services they offer, others regularly cross the line in their work. Recently, Duke University began asking on its college application whether students had received help with their application material. The question was added amid growing evidence that many private college counselors are writing, or extensively editing, the personal essays of students. Elsewhere, admissions officers are growing adept at spotting overly "packaged" candidates. "One of the reasons we ask for a graded paper [in the college application] is that we can see a big difference in the quality of work that has been handed in for a course and what has been polished up for a college essay," says Jane Brown, who oversees admissions at Mount Holyoke College, in western Massachusetts. "We look for disjuncture in the application—to see who has been packaged." Brown says about the applications process that "cheating is up and it's not on the part of poor students necessarily. It's students who've done well and feel pressure to keep up.... There is a tremendous stress around getting into brand-name schools."

Down the road from Mount Holyoke, at ultra-competitive Amherst College, Dean of Admissions Tom Parker and his staff are also scrutinizing applications for signs of cheating. "Where we earn our money is when the evidence is mixed or contradictory—a bad essay with high scores or a terrific essay and a verbal score in the 500 or 600s." Parker, who has been in the admissions business for twenty years, says things are worse now than ever before. "There is a current cultural obsession with getting into a particular set of colleges, that somehow then your life will be taken care of, or if you don't the opposite will happen.... The hype and anxiety have grown to a fever pitch. It's also spreading around the country to places where it didn't exist. It's anxious parents. It's a changing economy."

The unethical help that high-paid private college counselors provide to high school students is paralleled by the rising problem of private tutors who do students' homework. The private tutoring industry has exploded in recent years. More and more wealthy parents already shelling out $20,000 a year for private-school tuition are also spending thousands of dollars on top of that for high-priced tutors. There are no licensing requirements for tutors, no ethical code of conduct, and no accountability to anyone except to the parents who write the checks. Tutors know that their job hinges on getting results. "I have been asked to edit papers, and even write or rewrite sections of them, as well as to complete homework and to do research for students," says one woman, who worked as a tutor in New York to kids attending elite private schools. "There have been times when I have refused to actually complete homework for a student when the student tried to insist on it, and have had another tutor hired to do that work in my place. There have also been times when I have worked on assignments for students against my better judgment because I wanted to keep my job." This former tutor came to see the transgressions of her trade as part of a broader pattern. "Parents help their children to cheat while they're in high school and then donate money and make phone calls to board members to help their kids get into college." Parents believe that "everyone" gets aggressive private tutoring for their kids, the woman added, and "feel that since everyone is doing it, their child would be at a terrible disadvantage if they didn't."

An English teacher at one of New York's most exclusive prep academies thinks that this problem is so pervasive that it corrupts the entire academic process. "Tutors write a lot of the kids' papers," she says. "The kids are so heavily tutored sometimes it's hard to tell what is their work and what isn't." The teacher relates an instance where she pointed out an idea in a student's paper that seemed to make no sense. "My tutor says that's right," the student replied confidently, as if she were invoking wisdom imparted to her by a senior consultant at McKinsey. To combat tutor-assisted cheating, and other common forms of cheating at the school, the teacher has turned to assigning more in-class writing work. She feels that she can't turn to parents for help in combating cheating, since the parents are part of the problem. "The parents' attitude is generally: 'Whatever gets you the grades, you should do. We don't care.'"

Many parents are going even further these days to help their kids. In a new trend, some parents are conspiring with doctors to manipulate the rules around disability to win extra time on the SATs for their perfectly capable child.

Cheating around disability rules is a delicate topic. The federal government bestowed official recognition on clinical learning disabilities in legislation passed by Congress in 1969. Within a single year, more than one million children had been designated learning disabled. The number of learning-disabled children—and the controversy surrounding their special status—has been growing ever since. By the mid-1990s, school districts across the country recognized 2.33 million students with learning disabilities. Children in the category soon accounted for more than half of all students with disabilities.[11]

While education officials in all fifty states use the term "specific learning disabilities" (SLD) to describe the problem of learning-disabled children, there is wide discrepancy in how this condition is defined and how students are granted SLD status. In some states, the diagnosis and handling of SLD students is narrowly regulated by law. Others provide little or no guidance to local school districts and individual schools. The result is widespread confusion that has nurtured controversy and opened the door to abuses of disability rules around the SATs.[12]

Educational Testing Services, which administers the SATs, has long permitted students with learning disabilities to have more time on the SATs—as much as double what is allowed for other students who take the three-hour test. ETS previously flagged the SAT scores of learning-disabled students who were granted the extra time, so that college admissions officers had this information. ETS dropped this practice a few years ago as a result of a 1999 lawsuit. Now the scores of students who take the SAT with extra time are not differentiated in any way. The change made sense and was hailed by disability advocates who tell horror stories of discrimination against disabled kids.

But well-to-do parents have been quick to pick up on the huge opportunity that now exists to manipulate the system. In Westchester County, Dr. Jeanne Dietrich, a psychologist, has noted a spike in the number of parents seeking a diagnosis of learning disability for their college-bound child. Some parents openly tell Dietrich that their child had done poorly on the SAT and press for a quick diagnosis in order to meet the deadline to retake the tests. Another Westchester psychologist, Dr. Dana Luck, commented: "More and more people are asking legitimately.... But more and more are also asking because, why not ask? It's part of our culture that every point matters, so they're looking for any kind of edge."

This new type of cheating is not affordable to everyone. Luck charges $2,500 for an examination and $250 an hour to lobby school and ETS officials to grant SLD status to one of her patients.[13] Of some 30,000 students nationwide who are granted such status every year, a disproportionately larger number come from wealthy communities. Meanwhile, numerous poorer kids who truly deserve to have disability status don't get it because their parents can't afford the diagnosis. Along with private tutors who cross the line and college counselors who package kids, twisting disability rules is one more way for parents to give their kids every advantage. "I think it's the culture," comments a disability activist. "It's the mentality of aggressive, competitive parents who are playing the system against the kids who really need it. It's truly sad for the kids who have issues." At Amherst, Tom Parker is disheartened by this cheating, along with so much else that he sees on his job. "The unflagging is a terrible problem. What saddens me about this is that it was an honorable thing that the College Board did.... But in the current atmosphere if you open the door a crack you have 5,000 people who want to manipulate it."

The new abuses around learning disabilities have a corrosive effect in academic environments already beset with ethical problems. The mother of a Horace Mann senior with mild dyslexia—but no formal SLD status—relates the outrage felt by her son about another student who manipulated the system. "Everybody knew this kid got the right tutors and extra time—that he cheated his way through school with a false LD diagnosis.... They grew up with him, and all of a sudden in high school he was getting extra time. This kid got in early to Penn and the other kids who played it straight were devastated. My daughter said to my son, 'You didn't play it right.'... What does this teach them ethically, because even though you're cheating your way, you're winning?"

The mother, who sent two of her children to Horace Mann, says that the problem reflects collusion between doctors, parents, kids, and school administrators. The kids have become adept at conning the system, too, by acting in ways that can secure them a disability status. "Years ago it was a stigma," she says about disability labeling. "Today it's another way to play the system for the people who know how to do it."

As THE RANKS of the affluent have swelled over the past two decades, so have the number of kids who receive every advantage in their education. The growing competition, in turn, has compelled more parents to spend more money and cut more corners in an effort to give their children an extra edge. Nothing less than an academic arms races is unfolding within the upper tiers of U.S. society. Yet even the most heroic—or sleazy—efforts don't guarantee a superior edge. Applications to the top schools reached their highest level ever in the late 1990s. In 1999–2000, the eight colleges of the Ivy League received 121,948 applications—and rejected 80 percent of them. In 1999, the freshman class that enrolled at Brown University was culled from a vast deluge of applicants in which 3,500 applicants had been one of the top five members of their graduating classes. In the same year, Tufts University rejected one in three valedictorians who applied, as well as a number of applicants with perfect 1600 scores on the SAT.[14] College admissions directors at the best schools talk about the immense challenge of winnowing down large applicant pools filled with one perfect candidate after the other.

Many people scoff at the importance attached to name-brand schools, and it's easy to condemn the less savory motives of parents who want a Harvard kid. But the reality is that in a winner-take-all economy, and a society increasingly obsessed with "branding," a degree from a prestigious college matters more than ever. For example, many recruiters for America's best companies focus their search for entry-level professionals exclusively on the top schools in the nation, and for good reason. As any headhunter will explain, hiring personnel is extremely time consuming and fraught with risk. Because so many hires do not work out, as many as half in some settings, employers are essentially playing a numbers game; the higher the ratio of good hires to bad hires, the less time and money that gets wasted. These simple facts drive employers to focus on what they think are the most promising pools of labor. While there's no guarantee that, individually, students from top universities will be good hires, the risk of bad hires is judged to be lower in the aggregate. A kid hired straight out of Harvard might start every day with a bong hit and smoke a joint at lunch, while a new hire from Boston University might work like a demon until eleven every night. Even so, when an employer hires a Harvard grad, they believe that the odds are working in their favor.

The hiring practices of a company like Microsoft show how brand-name degrees can translate into a gilded rise to the Winning Class. During the 1980s, when Microsoft was still small, it exclusively recruited from fifteen top universities, including Harvard, Yale, Carnegie Mellon, and MIT. Brilliant geeks from the rest of America's colleges never had a chance to get in on the ground floor at Microsoft and land the generous stocks options that were part of the compensation packages for new programmers. Many of Microsoft's early employees went on to become multimillionaires.[15]

McKinsey & Company is another example of a leading firm that focuses its recruiting mainly on a pool of name-brand schools. The blue-chip consulting firm makes its money by convincing clients that the smartest people in the world will be coming in to solve their problems or improve their organization. McKinsey's leaders have long felt that this claim will be more believable if the firm is filled with Ivy grads and so McKinsey concentrates its recruiting efforts on these schools. A college grad or freshly minted Ph.D. who is tapped to join McKinsey has an opportunity that is simply not available to similarly brilliant people from other schools—namely, the opportunity to make partner at the immensely profitable firm and become a millionaire quite early on in life. This same kind of unique chance awaits law students who land jobs at white-shoe law firms like Cravath, Swaine & Moore, which also focuses its recruiting strictly on top name-brand schools.

In one study conducted by economists Robert Frank and Philip Cook, over a third of corporate recruiters indicated that they were focusing more attention on top-rated universities. Elite firms were most likely to be narrowly focused in their recruiting efforts.[16] Anecdotal evidence suggests that this discrimination occurs nearly every day at every kind of organization in America. Gatekeepers for the best corporations, government offices, law firms, publishing houses, film production companies, nonprofit organizations, and media outlets all gravitate toward applicants with name-brand degrees.

The more general trend of rising income gaps across the workforce has also increased the stakes of education. "Over an adult's working life, high school graduates can expect, on average, to earn $1.2 million," reports the Census Bureau. "Those with a bachelor's degree, $2.1 million; and people with a master's degree, $2.5 million. People with doctoral ($3.4 million) and professional degrees ($4.4 million) do even better." These earnings gaps have increased steadily over the past few decades. In 1975, workers with advanced degrees earned 1.8 times as much as high school graduates. This gap increased to 2.6 times in 1999.[17]

Rising education costs place a further premium on doing well academically. If you're a straight-A student, you have a better shot at various scholarships and awards that can defray the cost of your education—or make it possible for you to afford college at all. In turn, the heavy debts that more young people graduate with make landing a good-paying job all the more crucial. These financial challenges have become greater in recent years, thanks to reduced government support for higher education. Tuition and fees at private and public universities have more than doubled in the past twenty years, outstripping the increase in various subsidies for higher education. More young people are going to college now than ever before, but since the late 1970s the gap in college attendance between low-income kids and wealthier young people has actually been growing—even as everyone acknowledges that more education is needed to make it in the information age.[18]

Young people understand all of this. "Students are remarkably sophisticated about these matters," write Frank and Cook, who argued that trends in education epitomize the winner-take-all society. "If access to the top jobs depends more and more on educational credentials, we would expect them to do everything in their power to improve their credentials, and indeed they have."[19]

We might also expect their parents to do everything possible. Given the staggering rewards and penalties now at stake in the battle for advancement, it's no surprise that parents will pay any price and break any rule to make sure that their child has every advantage—from the first days of nursery school until that imagined moment when the family SUV rolls up to a Harvard dorm on enrollment day. Or, for parents with lesser dreams and fewer means, that imagined moment when their child wins a full scholarship to a nearby state school.

It's no surprise, either, that once a student is in college, he or she will sense that the real competition has just begun.

"College is only one of many destinations in the fast lane," observed Harvard's dean of admissions and his colleagues. "The accumulation of 'credentials' simply continues to intensify as the stakes increase. The 'right' graduate school looms after college, and the 'right' sequence of jobs is next. Such attainments make it possible to live in the 'right' kinds of communities and begin the job of bringing up the following generation, one that might need to vault to even greater heights."[20]

How to assure smooth forward movement in the higher-education parts of this arduous journey? Be prepared to cut corners early and often.

Cheating by college students has long been a problem. In 1931, Dean Clarence W. Mendell of Yale declared the problem of cheating at the school to be "so prevalent as to demand instant and sweeping measures of reform."[21] Hundreds of studies have been conducted over the past eighty years that look at why, when, and how college students cheat on their academic work. In a 1938 survey, a majority of students who indicated they thought it was "right to cheat" justified cheating on the grounds that "it gives one a chance to keep up with those who do cheat."[22] A 1941 survey of college students discovered a dramatically higher incidence of cheating among members of fraternities—a jump attributed to the requirement that members maintain a high grade point average.[23] The academic study of cheating grew after a high-profile 1951 scandal, in which nearly ninety cadets were dismissed from the United States Military Academy for taking part in a conspiracy to get test questions in advance.[24]

In 1964, William Bowers published *Student Dishonesty and Its Control in College*, the most authoritative study up to that time on academic cheating. Based on surveys of more than 5,000 students at ninety-nine colleges and universities, Bowers concluded that three quarters of all students had engaged in some kind of cheating, and he drew a variety of other conclusions from his data. He found that students who ranked social and professional aspirations at the top of their list of college priorities tended to be more likely to cheat than students who saw college as a training ground for moral and intellectual development. He also found that in "most cases there is no difference in the likelihood of cheating among students of different social backgrounds who were attending the same type of school."[25]

Scores of scholars followed in Bowers's footsteps after the publication of his seminal study, confirming the high level of student cheating. Different methodologies were developed to document and explain academic cheating. The research also moved beyond college students, examining high school and graduate students. To overcome the notorious problem of surveys, namely that they depend on "self-reporting," some researchers concocted controlled experiments where students were given an opportunity to cheat and then were carefully observed.

These days, if the education establishment had a chief detective, it would be Donald McCabe. McCabe is a professor of management at Rutgers University in Newark, New Jersey, and the founder of the Center for Academic Integrity. Years ago, like many professors in academia, McCabe was shocked to find out how much cheating went on among college students. Unlike most professors, who do little about such cheating, McCabe decided to take action. He began researching the problem in the early 1990s with rigorous surveys involving thousands of students. He also founded the center and became its first president. After a decade of research, including six major studies, McCabe is without question the leading national authority on cheating among high school and college students. McCabe's surveys at dozens of college campuses have revealed overall levels of cheating similar to what Bowers found in the 1960s—with roughly three quarters of students confessing to some kind of cheating—but McCabe suggests that today's college students are more likely to be engaged in serious cheating. McCabe has also documented increases of cheating in the 1990s of between 30 and 35 percent.

Why college students cheat remains a complex and disputed matter. Students cite a wide range of factors in explaining their cheating, including time pressures, the ease of cheating via the Internet, and the tolerance of cheating by faculty. McCabe sets this array of explanations within the broader context of today's high-stakes academic environment. Writing with two colleagues in 2001, McCabe commented: "With increasing competition for the most desired positions in the job market and for the few coveted places available at the nation's leading business, law, and medical schools, today's undergraduates experience considerable pressure to do well. Research shows that all too often these pressures lead to decisions to engage in various forms of academic dishonesty." McCabe also suggests that a tipping point has been passed in many academic environments. "Students who might otherwise complete their work honestly...convince themselves they cannot afford to be disadvantaged by students who cheat and go unreported or unpunished. Although many find it distasteful, they too begin to cheat to 'level the playing field.'"[26]

McCabe and his colleagues have explored cheating not just through surveys but through campus focus groups where students are encouraged to discuss their reasons for cheating. Students indicate a deep cynicism about what it takes to make it in America. "The world isn't fair and sometimes to get where you want you have to sacrifice some integrity," said one student. Another commented, "I figure a (small or large) percentage of the student body is already cheating in order to improve their grade, so I might as well cheat once in a while to help myself. I also believe that a portion of 'successful' people in today's world have cheated in their life from time to time, and they are successful because they have been smart enough to avoid getting caught."

The choice between being a winner or a loser in an economy filled with inequities seems stark and frightening to many college students. Says one student: "Grades are the most important things which judge whether you go to medical school or to work as a janitor."[27]

Academic dishonesty increasingly continues after college in graduate programs. Cheating among graduate students has been far less well researched than cheating at the undergraduate level, but a number of studies suggest that graduate school cheating may be as serious a problem as undergraduate cheating—especially in law, business, and medical schools, which are all training students to perform critical tasks and are nurturing the future leaders of our society.[28]

What becomes of the many students who cheat their way through school? Well, according to some scholarly research, young people who cheat in academics are more likely to cheat in other environments, such as workplace or business situations, and on taxes.[29]

Business students are among those with the worst attitudes toward cheating, and those most likely to bring lax ethics into their professional lives. A 2001 study of 1,000 business students on six campuses found that "students who engaged in dishonest behavior in their college classes were more likely to engage in dishonest behavior on the job."[30] With up to a quarter of college students typically choosing to major in business or a related field, and over 100,000 MBAs graduating annually, widespread cheating among business students is not an insignificant problem.

CHEATING IN POST-ACADEMIC life often begins during the creative sculpting of the all-important résumé and the holy quest for the right job. The job search is the culmination of years of sacrifice and toil. For many, it is a moment of truth: Will you succeed or will you end up working at Blockbuster?

During the boom years of the 1990s, the stakes for elite young job seekers were higher than ever before. Join the right dotcom start-up, with a generous package of stock options, and you could find yourself transformed into a centamillionaire in a few short years. Get passed over for another candidate, and that gilded dream vanishes into thin air, replaced by the dreary prospect of actually working your way up in the world. Today, in the aftermath of the boom, the stakes are also high—namely, basic survival. Over two million jobs disappeared in the U.S. between 2001 and 2003, with some of the most competitive and lucrative industries getting hit the hardest. Stories abound of highly educated young professionals working in sales jobs or not working at all.

With the stakes of job hunting now so high in both good times and bad, it should come as no surprise that more job seekers misrepresent their credentials. The American résumé, in fact, is right up there with lawyers' time sheets and corporate earnings statements as among the most misleading documents around.

Many people start lying on their résumés while in school and continue to do so throughout their careers. A 1997 study by a company that does preemployment screening found that 95 percent of college-age respondents were willing to lie in order to get a job—and that 41 percent of the students had already done so. Veterans of employment placement firms and human resource offices say that while résumé padding has always been a problem, it's reached crisis proportions in recent years. A review of 2.6 million job applications in 2002, by a national firm that conducts background checks, revealed that 44 percent contained at least some lies. Likewise, 41 percent of applications reviewed by a New Jersey–based verification firm contained information about education that was contradicted by the records of named institutions.[31]

In another large survey, HireRight, an Internet company that does background checks, found that 80 percent of all résumés were misleading—and a fifth included fabricated degrees. Since 1995, Jude Werra, a headhunter based in Wisconsin, has published what he calls the Liar's Index, which is based upon the percentage of résumés that he reviews that refer to bogus degrees. The Liar's Index reached its peak in the first half of 2000, with 23.3 percent of the résumés failing the accuracy test.[32] While Werra's data suggests that the greed of boom times brings out more lying than the anxiety of bad times, other headhunters disagree. "Since the bubble burst there is far more supply than demand," says Arnold Huberman, who runs his own search firm in Manhattan specializing in public relations jobs. "It's a much more competitive landscape. Therefore if someone comes in and shows us two graduate degrees, we'll check that."

Young people right out of school or still early in their careers are likely to be most insecure about their credentials. But résumé padding also goes on at the very top of the employment food chain. Christian & Timbers, a search firm that handles applications for CEOs and other top executives, has reported that up to a quarter of candidates provide misleading information.[33]

Many CV cheaters go far before their lies catch up with them. In 2002, Ronald Zarrella, the CEO of Bausch & Lomb, one of America's largest pharmaceutical companies, admitted that he did not have an M.B.A. from NYU, as he had long contended. Zarrella initially claimed that he hadn't proofread his official bio carefully enough. Later, when it was pointed that the same "typo" had been repeated in numerous news releases, including during his previous job, he came clean about his "lapse in judgment." Zarrella stayed on as CEO but was docked one year's bonus pay. The lie cost him over $1 million—a pittance, given that his fake credentials had helped him make many millions of dollars before he was unmasked.[34]

A year earlier, George O'Leary was hired as Notre Dame's football coach in a seven-figure deal—and fired shortly afterward when it was learned that he had lied about having a master's degree from NYU. (What is it about NYU?) O'Leary's brother Tom offered a spirited defense of him to a *Sports Illustrated* reporter: "Is anyone trying to tell me that résumés are truthful? In the America we live in, the willingness to lie on a résumé is an indication of how much you want the job."[35]

Olympic head Sandra Baldwin is another high-profile figure recently ousted from a lucrative job after the truth caught up with her. Her official résumé at the U.S. Olympic Committee claimed that she had graduated from the University of Colorado in 1962 and then obtained her Ph.D. in English from Arizona State in 1967. In fact, Baldwin had only earned an M.A. in English at Arizona State in 1969. "I knew how important education had been to my folks," she told a *New York Times* reporter, in recounting the difficulties of surviving the death of both parents when she was eighteen years old. "This put me on the course of not quitting." Baldwin even cited the name of her nonexistent dissertation—"Neo-Classical Backgrounds of Nathaniel Hawthorne's Aesthetics." In the same interview she asked: "What do you do with a Ph.D. in English?"[36]

One might wonder just how, exactly, anyone gets hired to a top management job with a fake educational degree. Even the laziest human resources bureaucrat or executive recruiter should be able to find time to make a five-minute phone call to a university records office.

Randy Neal, a managing director at an executive search firm in Dallas, recalls the ease with which he uncovered the lies of a would-be company head. "We were doing a search for a CEO and a candidate presented his credentials," Neal remembers. "He had a pretty long list with pretty big success stories with some well-known and some less well-known companies. The guy was an extremely articulate talker. I met him at the airport and he impressed me.... The client met the candidate and, like myself, was totally convinced that this was the guy." At that point, Neal started to do some digging, checking the man's background and references. "Right away we saw a red flag. A company that he had portrayed as a $200 million company really consisted of only two people. He tried to explain, but it was a complete falsehood. As we got checking further we started to find further inconsistencies.... Here is an example of a person who completely falsified his background."

Neal chuckles at the memory and starts in on another chilling yarn. "I had a case where a candidate had assumed someone else's identity." Like other headhunters, Neal has a quite a few of these stories. "We're talking executive-level positions," he says.

Neal and other professionals in the hiring business say that there are five or six common kinds of lies that appear on résumés.

Each has its own logic. People lie about their educational credentials for the obvious reason that better-educated people get better jobs and are paid more money. The income gaps related to credentials help explain the desperation reflected on résumés, which can be downright farcical at times. Monica Ronan, who hires for *TV Guide*, recalls checking out one job candidate's claim of a B.A. only to find that the college didn't exist at all. In cases like these, an applicant's stupidity may be a better reason for rejection than their dishonesty.

Beyond educational credentials, candidates also lie about how much money they made in past positions. "There are a lot of people who will stretch the truth a little bit—saying they made $200,000 instead of $180,000," Randy Neal says. Why? Because future compensation is almost always based upon past compensation. Other lies include length of job tenure (nobody wants to look flaky), reasons for dismissal (nobody wants to admit they were fired), and level of responsibility (everyone wants to seem more experienced than they are).

In 2002, shareholders in the software company Veritas—which means *truth* in Latin—saw the value of their holdings plummet by 20 percent when it was revealed that the company's chief financial officer, Kenneth Lonchar, had lied about having an M.B.A. from Stanford. Understandably, the market was a bit unnerved by the news that the same guy who signed off on earnings reports had also fabricated his résumé.

High-tech investors were even more rattled several years earlier when Lotus president Jeff Papows was ousted after it was learned that he had not only fabricated his academic record but also misrepresented his tae kwon do ranking and lied about past military service.

Papows, who led a $1.4 billion subsidiary of IBM, even falsely claimed to have been an orphan.

PERVASIVE DISHONESTY among students and job seekers is often treated as a puzzle by the media. Reporters parachute into schools to listen to tales of cheating, and then write stories that offer a muddle of explanations. Or, when some bigwig is found to have lied about his degrees, the media will ask a shrink to explain where the loose screw might be.

Yet maybe the real mystery is that there isn't *more* cheating by young people and job seekers. After all, the stakes are enormously high. The difference between getting into Harvard and getting into, say, Rutgers can easily shake out to several millions dollars over a lifetime. And the costs of being unemployed are greater now than in the past, with skimpier unemployment benefits and higher costs for necessities like housing and health care. More generally, the difference between good jobs and bad jobs is big and getting bigger in American society. Nobody wants to be caught on the wrong side of the widening chasm between the haves and have-nots. Cheating is one way not to be left behind.

CHAPTER EIGHT

Crime and No Punishment

WE LOVE TO PUNISH PEOPLE IN AMERICA—SOME PEOPLE, at least.

The United States is more punitive than any other advanced democratic society. We stand alone among such nations in putting people to death. We have "three strikes" policies that can send people to jail for life for petty theft. We are uniquely tough with the poor and unemployed, cutting off benefits to the jobless whether the economy has improved or not. We mete out long prison terms for drug offenses that are treated as personal health problems in Western Europe or Canada. We expel children from our schools for misbehavior under "zero tolerance" policies. For a while, we even had a Speaker of the House (Newt Gingrich) who advocated forcing unwed mothers to give up their children to orphanages.

Toughness runs deep in the veins of American culture. We imagine ourselves as a country where everyone is responsible for themselves and if you don't pull yourself up by your bootstraps, something must be wrong with you. The linguist George Lakoff has called this "strict father morality"—a code stressing that personal improvement and social order are best achieved by tough rules along with swift punishment to enforce those rules. "The strict father wants his kids to become disciplined, pursue the self-interest, and become self-reliant and be good people because they are disciplined," explains Lakoff.[1]

Strict father morality jibes easily with laissez-faire ideas about the moral health that flows from competition, and it bolsters the conviction that government shouldn't help people through social programs. Over the past quarter century, get-tough notions have spread in step with other values associated with the market and have helped to usher in big changes in our politics. Conservatives brought the curtain down on an era of liberal dominance in large part by caricaturing liberals as breeders of social pathology—as the indulgent political parents to violent criminals, lazy welfare moms, and thieving drug addicts. To save the day, and America's soul, conservatives have promoted a society that is harsher, more punitive, and less forgiving—assuring us that these steps are for everyone's own good.

And yet this toughness does not extend to everyone. While a punitive morality increasingly governs the lives of those in America with little wealth or power—the poor, minorities, immigrants—better-off Americans are actually coddled and nurtured more than ever, whatever their sins. Big-time white-collar criminals are untouched by prosecutors, SAT cheats go to Harvard, famous wrongdoers are feted by the media and paid six figures for their "confessions," ex-cons emerge from country-club prisons with healthy tans and go on to make millions of dollars through insider business dealings.

These second chances aren't doled out in Appalachia or Harlem or Roxbury. Indulgent morality thrives the most where the prerogatives of class privilege rule, big profits are at stake, and government enforcers have been disarmed.

And that is a lot of places.

A FEW YEARS BACK the co-valedictorian of Taylor Allderdice, the most prestigious public high school in Pittsburgh, was booed by some of his classmates at graduation. Flyers circulated in the audience raising "doubts" about the co-valedictorian's "academic integrity." The star student was from a well-to-do family. He had been enrolled in Allderdice's Center for Advanced Studies, an academic track for the best of the best. There, amid feverish competition for Ivy League admission, he was allegedly one of three students who consulted a dictionary during the SAT. When another student reported the violation, an avalanche of cheating stories tumbled forward from honest students who had, till then, kept their mouths shuts about rampant cheating at Allderdice. Students were said to have offered money for homework and to have used their calculators to store prohibited information during math and science exams. Test stealing had allegedly become so brazen and widespread that a "repeat cheater" did not hesitate to run down the hall brandishing his prize, shouting, "I've got it!" One student, Lianne Mantione, commented about Allderdice: "Cheating has just been a way of life."

What came of these troubling allegations? Nothing. Neither Taylor Allderdice nor the Educational Testing Service acted to penalize the alleged offenders in the SAT case, citing conflicting accounts of what happened. The so-called "chief cheater," in fact, not only was made co-valedictorian of his class but headed off to a top college. The girl who initially reported the incident found herself harassed in the hallways. Her family received obscene phone calls in the middle of the night.[2]

The episode was deeply disillusioning to those students who played by the rules, but it was not surprising. People who have gone to Taylor Allderdice remember it as the kind of school where parents call up teachers to hassle them about their kids' grades—and get the principal to hassle the teachers. It was also not uncommon at Allderdice for the principal to change student grades without the consent of teachers.

Graduates of America's top high schools and colleges tell many similar stories about permissive environments where cheaters thrive. For example, Ronah Sandan, a recent graduate of Tenafly High School, describes a school where cheaters had nothing to fear and much to gain. Many of the students at Tenafly have wealthy parents; the kids drive new SUVs to school, chat away on their cell phones between class, and wear the latest designer clothes. To ensure their future position in the Winning Class, these high school students will do nearly anything to succeed. Cheating was rampant in Sandan's advanced placement classes, filled with the top students in the school. "They had elaborate schemes for cheating in every class," Sandan says. "These were people who could have done well even if they didn't cheat." Yet despite widespread knowledge among teachers of the cheating, "there was so little enforcement." In one case, a student was caught cheating on the SAT by a proctor and suffered no penalty at all, besides having to take the test over. The student went on to attend Wharton. Other known cheaters also glided upward. "One of three guys who was known to be one of the biggest cheaters went on to Harvard," says Sandan. "It was an appalling situation. . . . It was a cutthroat, fend-for-yourself kind of thing. People didn't think about the fact that they would take someone's rightful place, because everybody was doing it, and it was so easy that if you didn't do it, you were a sucker."

Sandan didn't cheat and didn't get into an Ivy League school, or any of her top-choice schools. She's now at Rutgers University, where she says the faculty are even more tolerant of cheating than they were at Tenafly. "It's very easy to cheat and most professors don't make much of a fuss about it, or make it clear there would be any repercussions."

Most academic cheating does, in fact, go unpunished. A consistent finding of the research on academic cheating is that there are few consequences for those suspected of cheating. In a 1999 survey of 1,000 faculty at twenty-one colleges, a third of professors said they were aware of cheating in their classes but didn't stop it. Likewise, in an earlier survey of student-affairs administrators in colleges across the United States, 60 percent reported that faculty at their schools tended to handle incidents of cheating independently and not subject student violations to formal disciplinary action.[3] Many professors would rather let cheaters slide than take on the bureaucratic hassles of pursuing disciplinary actions. Others are afraid of lawsuits filed by the parents of cheaters. In a 1999 survey by the *American School Board Journal*, roughly half of teachers said that the threat of litigation discourages them from punishing student cheaters.[4] When cases are brought against students, the wealth of their parents can help neutralize the fallout in other ways. Stuart Gilman, president of the Ethics Resource Center in Washington, D.C., recalls an episode from his days as a college professor. "I had a student who handed in an unsatisfactory paper. The dean then begged me to give him a ten-day extension. Then he handed it in (this is the day before graduation) and it was plagiarized. It had nothing to do with the topic, and I felt like I had read it somewhere. It turns out it was an article in the *American Political Science Review*. I failed him. The course was a departmental requirement to graduate in political science. I was at graduation and I saw him there. I thought, 'Oh that's nice, they let him come.' But then he came over to me and said he was actually graduating even though he had failed my course, because they had waived the requirement for him."

Gilman later learned from the dean why the student had been allowed to graduate: "The student's father had just given a million-dollar contribution toward a new building on campus."

In many elite academic settings the presumption of innocence is given to students who, along with their parents, are effectively an affluent "client" base—often with a multigenerational brand loyalty. In contrast, faculty are merely the hired help. When faced with the choice of keeping its clients happy or standing behind the hired help, schools often make the same choice that any self-preserving business might make.

In a 2000 incident at Dartmouth College, a visiting instructor accused students in his introductory computer science course of cheating. An anonymous telephone call tipped off Rex Dwyer that students had visited his personal Web page and downloaded answers to a class assignment. Though the cheating was initiated by members of the Gamma Delta Chi fraternity, access to Dwyer's Web site was shared widely. Eventually, Dwyer learned that his site had been accessed by thirty-two different campus computers, and that seventy-eight of his students had taken part in the cheating.

Dwyer filed formal charges with the university and ended up spending fifty hours documenting the case, writing numerous memos describing in detail his evidence of cheating, which included server Web logs. Yet in the end, even though investigators agreed that cheating had occurred in the class, Dartmouth's disciplinary body, the Committee on Standards, dismissed all the charges brought by Dwyer, saying it was too difficult to verify exactly who had cheated. Dwyer said that strong evidence did exist, and that he could have gotten much more precise evidence if he had been allowed to access other campus computer records—a request that Dartmouth denied. "More bizarre," he wrote, "is that the Committee's members apparently were less willing to acknowledge the truth than some of the accused."

Dwyer condemned the investigation as a "whitewash" aimed at avoiding a "public relations nightmare." Beyond his many hours of work on the case, Dwyer's reward for his commitment to academic integrity was to be publicly maligned for being a poor teacher and, some alleged, even tempting the students to cheat in the first place.[5]

Many top universities create a permissive environment around cheating by failing to institute tough honors codes. The research of Donald McCabe and others shows clearly that honor codes and/or a serious institutional commitment to academic integrity reduce cheating.[6] Still, many universities haven't enacted honor codes, or made such a commitment. Some seem to fear that doing so will attract negative attention. But there is also the issue of resources. Creating an honest academic culture on campus can't be done by just passing an honor code or putting flyers in students' mailboxes. It requires sustained and involved efforts, every year, by faculty and administrators. Given these downsides, it's a whole lot easier for a school's leaders to pretend that cheating is at most a minor problem.

Yale is a classic case of a school in denial. Yale first experienced a high-profile cheating scandal seventy years ago, and it has been rocked by periodic scandals ever since. But it has never instituted an honor code. "The reason they don't have such a firm policy at Yale, they say, is that they don't need it," says Andrea Spencer, a Yale senior. "They say if they made something like that, it would imply that they need it, so it would reflect badly on them. It's kind of ridiculous because they do need it. There is a lot of cheating here.... Most professors say they don't really look for it. That it doesn't happen much. But if you look the other way, of course you're not going to find it."

Cheating involving athletes is also widely tolerated at universities. School athletics departments are awash in scandal, including cash briberies to recruit high school athletes and flunking athletes who are still allowed to compete on the playing field. The case of a college athlete like Andre Johnson is a good illustration. Johnson is a wide receiver who plays for the Miami Hurricanes and turned in a star performance at the 2001 Rose Bowl. But in 2002 he was found to have cheated on an exam and also to have plagiarized a term paper. Following its internal rules, the university planned to suspend him for two full semesters. Then it changed its mind, downgrading the punishment to suspension for two semesters of summer school, a meaningless slap on the wrist.

Tom Petersen, the professor who brought Johnson's cheating to light, felt conflicted about turning in the star athlete and hoped that the incident could spark a dialogue on campus about the pressures that young athletes are often under. He wrote a set of proposals along these lines, but he was ignored by the school's administrators and castigated by others. "I was demonized for talking about a national, systemic problem with the way we exalt college football," Petersen said later. "There is fear and loathing among the faculty on football issues. Nobody wants to beat up on Andre Johnson."[7] The university was mum about why it changed Johnson's penalty. But it is safe to speculate that the risk Johnson's suspension posed to the Hurricanes' performance—and box-office and broadcasting sales—was simply too great.

The corruption of college athletics is pervasive and well known. Johnson's slap on the wrist is typical in a world where both coaches and college administrators are either breaking rules themselves or turning a blind eye when athletes break rules. Despite growing regulation over the past two decades by the National Collegiate Athletic Association (NCAA) and other bodies like the Southeastern Conference, rules are bent or ignored all the time. Cheating coaches and top athletes are let off the hook again and again. Schools are put on probation one year, only to violate the rules again the next year. Professors who try to enforce academic rules against athletes find themselves receiving hate mail and obscene phone calls. Some, like Linda Bensel-Meyers, a Renaissance scholar at the University of Tennessee, have even come to fear for their physical safety on campus after reporting academic fraud by athletes.[8]

College athletics has always been an intensely emotional world. But now it is also big business, which explains a lot. University stadiums that used to hold 20,000 people hold 90,000 today. Television rights have become more lucrative with the rise of cable and ESPN, and more viewers have attracted more advertising dollars and more corporate endorsement money. By the late 1990s, the NCAA was pocketing an estimated $270 million in annual income by selling the rights to television broadcasts and collecting dues and fees. In early 2003, the NCAA announced that CBS would pay $6 billion over eleven years for broadcasting rights covering twenty-two college sports. Winning teams can become a cash cow for universities, helping to cover the cost of expensive athletic programs and bringing a luster of prestige to the school. Losing teams can send athletic revenues plunging and lead to a falloff in donations from angry alums.[9]

The pressure to have winning teams is felt acutely by university presidents, whose job security depends on the goodwill of leading alums and other power brokers connected to universities. "A lot of people just care what happens on a Saturday afternoon and they put pressure on trustees and presidents that may be at odds with the direction the university should be headed," says

James Duderstadt, who as president of the University of Michigan for eight years learned "a great deal about corruption and commercialism in university sports." Duderstadt wrote a scathing book about his experience in fighting these trends. It portrays university presidents as the puppets of powerful alums focused on sports victories who work in close concert with savvy allies within the university, as well as outside commercial interests. "Many presidents don't have power because they report to governing boards of lay citizens who might have more interest in a winning team than in academics," Duderstadt says. "The boards are put in place to support the sports teams, to act on behalf of coaches. A lot of presidents keep a low profile and don't challenge it. They want to choose the ditch they are going to die in, and that's not the one." These dynamics mean that university sports coaches have wide latitude to do what it takes to produce winning teams. And this means that cheating is widely tolerated. Like other college presidents, Duderstadt found himself aware of these practices but felt unable to stop them.[10]

Top football and basketball coaches at big universities may have the most ethically dodgy jobs in America today. They are often the highest paid and most visible figures on a university campus, and they inhabit a world bracketed by fat carrots and unforgiving sticks. The earnings of top coaches can exceed several million dollars a year from a combination of seven-figure salaries, endorsement deals, hefty bonus payments by alumni "booster" groups, and lucrative speaking gigs. Coaches can become heroes and celebrities—as long as they keep winning. "If you don't win, you're not doing your job," commented University of Texas athletic director DeLoss Dodds. "And you don't survive."[11] To win and keep winning, coaches often leave their ethics at church. "There's a turn-it-around-quick mentality which encourages you to do whatever you need to do to turn the program around, which sometimes means cheating," explained Bill Battle, who was the head football coach at the University of Tennessee for six years.

The most common forms of cheating by coaches involve improper recruiting tactics and the flouting of rules governing the academic performance of students. For example, University of Georgia basketball coach Jim Harrick, Sr., was forced out of his job in 2003 after he was caught in an academic fraud case too serious to be swept under the rug. The case involved his son, Jim Harrick, Jr., who was an assistant coach under his father. The junior Harrick was accused of helping some of the team's players get through school with little work. He taught a class, Coaching Principles and Strategies of Basketball, in which everyone received A's, even students who never attended class or did any work.

After Jim Harrick, Sr., was fired many people wondered why he had been hired in the first place. This is a man, after all, who was accused of academic fraud and sexual harassment in his previous job at the University of Rhode Island. He had also been fired from a coaching job at UCLA for lying on expense reports. Yet apparently these past problems were judged insignificant in comparison to his outstanding coaching credentials—over a twenty-three-year career, Harrick had taken four different teams to NCAA tournaments. It was this record that impelled Georgia to hire Harrick at a $750,000-a-year salary to restore the school's lost stature in collegiate basketball.[12]

Harrick's involvement in academic fraud was not unusual. Coaches routinely help athletes shirk their studies and circumvent rules designed to ensure that this doesn't happen. "Less than 50 percent of football players and less than a third of basketball players will ever get a degree," says Duderstadt. "We have coaches making millions of dollars a year and advertisers and broadcasters making millions of dollars and the kids are being exploited. That's the most serious indictment."

The NCAA is supposed to vigilantly police the world of college athletics. It has at its disposal a tough array of sanctions, which have gotten fiercer since the 1970s. But like everyone else involved in college sports, NCAA officials have a strong incentive to look the other way when cheating occurs. The NCAA's broadcasting revenues depend on quality competition between strong teams with stellar athletes. Actually enforcing the rules often works at odds with this goal, since it can lead to the disqualification of top athletes or even entire teams. "The foxes are in control of the henhouse at the NCAA," comments Duderstadt, expressing a widely shared view.

Meanwhile, very little is done about the most common form of cheating in pro sports: drug use. While the NFL has a serious drug-testing policy, major league baseball—where the use of steroids and human growth hormones is raging out of control—has no mandatory drug testing. Nor does the NHL. "The current drug tests we have—only careless and stupid people flunk them," says Charles Yesalis, a leading authority on drugs in sports who teaches epidemiology at Penn State. "It's done for public relations, directed at naïve journalists and naïve fans." This means there is no realistic way to sanction players who are violating the rules on performance-enhancing drugs, except if those athletes voluntarily turn themselves in. To toughen things up would mean taking on the unions that represent players and also drawing attention to the problems of drugs in sports.

Nobody has an interest in doing this. The public certainly isn't clamoring for a crackdown—in one 2003 poll, for instance, only 16 percent of Americans said they thought drugs in baseball were a problem (compared to 40 percent concerned about pay gaps among players).[13] The teams and the leagues also aren't anxious to take action. "Unless the drugs hit the bottom line, unless it impacts them financially, nothing will ever be done," says Yesalis. "Many of these drugs work and it is the bigger-than-life athlete doing the bigger-than-life feat that has made sports the multibillion-dollar industry it is. The entertainment value is so great, the money is so great, and the fans don't care. That's why it will continue."

MANY CHEATERS ALSO GET a free ride outside the training grounds of academia and the rarefied sports realm. In particular, while a crackdown on street crime sent America's prison population surging to record levels in the '90s, the perpetrators of "suite crime" enjoyed the good life. A growing epidemic of white-collar crime attracted little notice, and no real response, until the collapse of Enron in late 2001.

The Association of Certified Fraud Examiners (ACFE) is located in Austin, Texas. The group's 28,000 members are spread throughout the country and many are skilled in the arcane specialty of forensic accounting. The subfield has grown rapidly in recent years. Conferences are held to exchange tips on how to unravel complex financial frauds and, in 1999, a group of academics founded the *Journal of Forensic Accounting*. Their timing couldn't have been better.

Fraud examiners are constantly struggling to keep up with sophisticated scams for embezzlement, padding expenses, hiding money overseas, and so on. These are the experts that get called when something doesn't seem right with a company's books or when an employee absconds with company cash. Such calls have been going up in recent years.

ACFE is the most authoritative source in the United States for data on "occupational fraud." In 2002, the group estimated the annual total losses from workplace fraud at $600 billion, or 6 percent of GDP. "It's a huge problem," the association's general counsel, John Warren, has said. "It's easily the most costly form of economic crime in the world." The 2002 survey of ACFE members found that while lower-level employees committed more acts of fraud, the misdeeds of those higher up the food chain were far more costly. For example, the median cost of financial statement fraud, one of most sophisticated forms of fraud, was $4.2 million. "The most costly frauds are committed by well-educated senior male executives," the report found.[14]

What happens to the professionals who perpetrate these inside jobs? Often not very much. Most fraud examiners reported in the survey that they felt that punishments were not sufficient in the cases they handled. A quarter of the cases were not even reported to law enforcement agencies, and 81 percent of the companies did not file a lawsuit to recover the money. Many fraud examiners believe that company managers lack the desire to prosecute criminal employees and that most don't do enough to prevent fraud. "Companies are very reluctant to air their dirty laundry in public," Joseph Wells has explained. Wells is a former FBI agent who is founder and chairman of ACFE. Most businesses handle the problem of a bad executive or lawyer or accountant with quiet discretion. "It doesn't do a lot for the depositors' confidence when they read in the newspaper that their bank just nabbed a crooked loan officer," said Wells. "As a result many businesses simply fire dishonest employees, who then are unwittingly hired by other companies."[15]

In theory, white-collar fraud and misconduct is policed not only by employers and law enforcement agencies but also by professional groups that are supposed to enforce codes of conduct among their members. Just about every kind of professional belongs to one of these groups, some of which have the power to revoke licenses. Doctors belong to the American Medical Association, lawyers belong to the American Bar Association, stockbrokers belong to the National Association of Securities Dealers, and so on.

These groups have a strong self-interest in policing their ranks, lest they forfeit public trust or end up getting more hassles from government regulators. Yet often that self-interest is sidelined and, in many associations, tolerance for cheaters is the rule rather than the exception.

State bar associations underscore this problem. These groups are the top cops within the legal profession, and each has its own set of disciplinary rules and procedures, as well as committees of lawyers that perform this work. But many state bars have far too few resources to do their job effectively and there is little national oversight of the groups—thanks to the influence of private law associations that consistently block reforms. In many cases, the disciplinary framework they operate within has not been evaluated or updated in decades. Some egregious forms of legal abuse, such as overbilling, are not even in the jurisdiction of state disciplinary groups. And, in nearly all states, people who lodge complaints against lawyers have no way to find out what is happening in their cases.

A 2002 study by HALT, a legal reform group, rated each state's disciplinary capacity using data provided by the state bars and the American Bar Association. The study, which used a report-card rating, found that thirty-nine states earned below a C, and none received an A. Most complaints in many states are not even investigated, much less result in punishment. Only 3 percent of investigations result in the sanctioning of lawyers, and just 1 percent end in disbarment. Punishments are typically meted out in secret, making them even less effective. "Because private sanctions are so lenient, the legal profession is not effectively deterred from engaging in unethical conduct and because these sanctions are held in secret, consumers are being deprived of valuable information about their lawyers' discipline history."[16]

Several commissions have called for reform of this flawed system, but nothing ever seems to change—however outrageous the lapses of discipline. For example, in the 1980s, dozens of lawyers were caught up in the savings and loan debacle and received hundreds of millions of dollars in penalties from the federal government, but not a single lawyer was disciplined by a state bar association for his or her role. More recently, it's become widely known that many lawyers played a key role in various of the big corporate scandals—yet there is little evidence that state bars will do any better at disciplining these rogues than it did after the savings and loan scandal.[17]

Given the low odds of being hassled by a state bar, it is a wonder that the Arkansas state bar undertook disciplinary proceedings against a lawyer whose main sin was that he lied about his personal life under oath. Then, again, Bill Clinton was no typical bad apple in the legal profession; he was unlucky enough to have many well-financed enemies.

State medical societies are on the front lines of policing the AMA's elaborate code of professional ethics. But these groups often do a poor job of helping either patients or doctors report ethical violations, and it is very difficult to find out what disciplinary actions are taken by these groups against physicians. Public Citizen, a consumer group in Washington, D.C., has commented that state medical societies "cannot prevent a doctor from practicing, and their vested interest, in most cases, is to protect their members, not the public." Looking out for one's own kind is a natural impulse. Major hospitals and managed-care companies, which also have the power to discipline doctors, often do not do much better—although hard facts are difficult to come by on this subject since most of these entities keep their disciplinary records secret.[18]

The financial services industry is another arena where self-policing is a joke. "Your broker probably has a better chance of getting busted for public drunkenness at the Indy 500 than being nabbed for impropriety by the NASD," commented one veteran Wall Street journalist who examined the disciplinary record of the National Association of Securities Dealers. A huge number of client complaints are dealt with by arbitration and don't ever make it to NASD's disciplinary panel. And many of the complaints that do get to NASD do not result in disciplinary action. This means that a broker can repeatedly commit unethical acts and yet seem to have a clean record.[19]

Accounting is also a profession that has its own watchdogs charged with keeping the bookkeepers honest. And how successful has this self-policing been in recent years? Let's not even go there.

SOARING PROFESSIONAL CRIME in the last two decades has not been accompanied by a beefing up of the resources needed to prosecute these crimes. Even as many states went on a prison-building binge through the '80s and '90s, and even as the federal government provided billions of dollars in new funds to help localities hire more police officers, and even as federal spending on the "war on drugs" approached $20 billion a year under President Clinton, strapped investigators on the white-collar beat often found themselves with little choice but to let high-level wrongdoers off the hook.

The losing battle against securities fraud is a case in point. Wherever this battle was fought in the past decade—be it Silicon Valley or Wall Street—government investigators were hopelessly outgunned.

In the late 1990s, more money was being made—and stolen—in Silicon Valley than anyplace else on earth. Following Netscape's extraordinary IPO in 1995, hot IPOs came fast and furious. It seemed that anyone with a half-decent business plan could raise millions, and that any company with a half-baked technology product could become worth billions in the stock market. However, once companies were publicly traded, life got harder for high-tech executives as they sought to show profitability and live under the tyranny of "the Number," that is, their quarterly earnings. Vast personal fortunes depended on what this number was. Executives who met the expectations of Wall Street analysts, even if only for a year or so, could make tens of millions of dollars exercising stock options and then cash out of the company for some other excellent adventure. Those who issued earnings reports that didn't meet these expectations felt like they were "signing their own corporate death warrants," in the words of ACFE's Joseph Wells.

The lure of easy riches was a recipe for bad behavior. In a story line that has since become familiar, many Silicon Valley companies started cooking their books with a variety of scams to misrepresent earnings and pump the value of their stock. Prosecutors in northern California found themselves totally overwhelmed by the crime wave. Even simple acts of securities fraud, such as releasing false earnings reports, are extraordinarily complex to prosecute. Investigators must first prove that a crime actually occurred, which often requires connecting the dots in a vast sea of fragmentary evidence and interpreting legal statutes and case histories that can run into the thousands of pages. Then they must prove intent and individual culpability—no easy task. High-ranking corporate officers often plead innocence on the grounds of "professional reliance": Because they depend upon the performance of a hierarchy of employees, accountants, and lawyers, they claim that they are immune from the charge of criminal intent.[20] The prosecutorial tasks of simply establishing strong grounds for indictment are labor intensive in the extreme. Equally daunting can be the job of convincing a jury of the case's merits in the face of counterinterpretations of every last shred of mind-numbingly complex evidence.

In his book *Infectious Greed,* regulatory expert Frank Partnoy explains that before the big corporate scandals in 2001 and 2002, there had been plenty of major financial crimes during the 1990s—including one that led to the bankruptcy of Orange County in 1994. Most of these crimes went almost entirely unpunished. The cases were too complicated and, writes Partnoy, "the defendants had plausible arguments about why what they had done was legal or fell outside the scope of existing law. Prosecutors did not bring many criminal cases, and they lost when they did."[21]

This pattern was repeated as the technology boom of the 1990s gathered steam. Despite a growing number of complaints alleging financial misfeasance by Silicon Valley companies, few new resources were added to combat the problem. By 1998, the Justice Department had exactly one assistant U.S. attorney working full-time in San Francisco on investment fraud, a profoundly beleaguered young prosecutor named Robert Crowe. Many criminal referrals from the SEC and FBI were not prosecuted. Crowe already had a huge backlog of Silicon Valley cases and was barely able to successfully bring to trial the few cases that he did go after. At one point, when he was preparing a securities fraud case against California Micro Devices, Crowe found himself deluged with 600,000 pages of documents and yet his office lacked a clerk to help sort and photocopy the materials. "There was no secretary, no paralegal, no resources I could count on from the office—nothing," Crowe said later. Facing a trial against Cal Micro's expensive team of lawyers and the army of paralegals at their disposal, Crowe simply was not able to go through all the documents. His plight reflected the fact that white-collar financial crimes were simply not a priority of a U.S. Justice Department that spent much of the 1990s worried about drugs, street crime, and terrorism.[22]

A similar story played out on the other side of the country, where prosecutors charged with policing a booming Wall Street in the late '90s were also outgunned.

The home of the Securities and Commodities Fraud Task Force, Southern District of New York, is on the third floor of a federal building in lower Manhattan not far from Wall Street. It is the largest U.S. government law enforcement team focused exclusively on crimes related to the stock market, including insider trading, false earnings reports, and other kinds of deception. The office became famous in the 1980s, when it prosecuted the insider trading scandals involving Ivan Boesky and Michael Milken.

In 2002, as scandals engulfed corporate America, the task force had only twenty-five lawyers. Assistant U.S. attorneys did their own copying. Numerous cases were not pursued because of a lack of resources. "We've got too many crooks and not enough cops," one federal regulator told *Fortune* magazine. "We could fill Riker's Island if we had the resources."[23]

It is a testament to the office's limitations that it fell to a then-obscure state official, New York State attorney general Eliot Spitzer, to pursue the massive conflict-of-interest scams involving stock analysts at top Wall Street firms. The rotten ethics of Henry Blodget, Jack Grubman, and others were no secret to those on the Street; all that was needed to bust open the case was an investigative staff that had the resources, and interest, to sift through a few thousand subpoenaed e-mails. Spitzer stepped into history only because the Justice Department's top watchdog on Wall Street was not up to the job.

In the end, Spitzer himself was not prepared to prosecute a hard-hitting securities fraud case against either the individual analysts who misled investors or the big investment banking firms. For all the tough talk that made him famous, Spitzer was smart enough to see that criminal prosecutions would have tied up his office for many years, and with uncertain results. Spitzer professed more interest in getting firms to change their internal practices than in getting convictions. His investigation led to investment banks adopting a host of new guidelines and paying a $1.4 billion settlement. Blodget and Grubman paid individual settlements that represented only a small percentage of their ill-gotten earnings. Blodget, who made $12 million in 2001 alone, only had to pay

$4 million; Grubman, who made a reported $20 million a year hyping worthless telecom stocks, and whose reported severance at Salomon Smith Barney had been $32 million, glided by with a $15 million tab. Neither man had to admit any wrongdoing as part of their settlement.

The most prestigious firms on Wall Street, along with the industry's most visible stock analysts, had lied to American investors on a massive scale. Hundreds of billions of dollars were lost on bad investments partly as a consequence of these multiple acts of securities fraud. Yet Spitzer's high-profile investigation did not result in a single individual or firm formally accepting legal responsibility. Nobody faced a judge or jury. Nobody saw the inside of a prison. And the most prominent villains in the saga remain multimillionaires today.

Would Blodget and Grubman do the same thing all over again if they had the chance? It's hard to see why not.

Eliot Spitzer was unusual in the entrepreneurial zeal that he brought to ferreting out wrongdoing in the securities industry. Prosecutors typically play a more passive role in policing Wall Street; most of their cases are generated by investigators at the SEC. While the Securities and Exchange Commission handles the lion's share of investigation into such malfeasance as fraud and insider trading, the agency lacks the prosecutorial authority to go after individual offenders with criminal charges. SEC investigators have been repeatedly frustrated over the past decade by sending cases over to the Justice Department only to see no action taken. Many white-collar crooks who were squarely in the crosshairs of government prosecutors have lived to steal another day. All told, SEC investigators referred 609 criminal cases to the Justice Department between 1992 and 2001. Federal prosecutors only followed up on 36 percent of the cases, and achieved convictions in 76 percent of these cases. Sixty percent of those convicted spent some time in prison—or a grand total of 87 offenders out of over 600 cases.[24]

There is a perverse irony in the idea that the SEC would overburden anyone with the results of its investigations, since the SEC itself is woefully understaffed. There are 17,000 publicly held companies in the United States and each must file earnings reports with the SEC, attesting under penalty of law that everything in these reports is true. An additional 9,000 mutual funds also regularly file reports with the SEC. At the same time, the SEC needs to worry about the behavior of 664,000 registered securities dealers working in roughly 5,300 brokerage firms with over 92,000 branch offices. These dealers work with as many as 50 million clients who are invested in a stock market where a billion shares of stocks and bonds change hands every day.

The SEC is responsible for ensuring the integrity of all of this activity and more. But its resources are paltry compared to its mandate. While the number of company reports filed with the SEC grew by 40 percent in the last half of the 1990s, staffing levels remained flat during this same period. In 2000, the SEC was only able to review 8 percent of financial statements filed by public companies—an alarming fact given that the agency knew that an epidemic of false earnings reports was under way. By some estimates, the SEC would need to quadruple the number of staff reviewing financial statements in order to review 30 to 35 percent of statements and achieve what it sees as an acceptable level of vigilance.[25]

THE 1990S MAY HAVE BEEN a decade where the mantra of "personal responsibility" was chanted by politicians of every stripe, but it was not a decade in which those who stole with a briefcase had much to fear. While scores of companies were fined or paid settlements as the result of white-collar criminal probes during the 1990s, the overwhelming majority of cases concluded with no individuals facing indictment, even in cases where multiple felonies had clearly been committed. This was true in cases against Sears, Merrill Lynch, Arthur Andersen (pre–Enron scandal), Waste Management, Sunbeam, Salomon Smith Barney, Johnson & Johnson, Bayer, and numerous other firms. Most recently, as of this writing, the federal government slapped AstraZeneca with a $355 million settlement for an illegal scheme to market a prostate-cancer drug. Yet while the company as a whole pleaded guilty to a single felony charge of health-care fraud, not a single individual was held responsible for what happened. Nor did the company bother sounding contrite about actions that defrauded taxpayers of tens of millions of dollars. "We disagree with the government on this, but to put it behind us, we are agreeing on a settlement today," said a company spokesperson.[26]

The rare white-collar criminals who actually did face a judge were invariably treated lightly. The zeal in the '80s and '90s to impose mandatory minimums and "three-strikes policies" never extended to white-collar crimes. For example, the executives at the agricultural giant Archer Daniels Midland who orchestrated a global price-fixing conspiracy that cost consumers $500 million faced maximum prison terms of only three years when they were sentenced in 1999. Even then the judge went easy on them, praising them as family men and community leaders, and giving them only two years in prison. In another case, Bruce J. Kingdon, who confessed to committing serious fraud at Bankers Trust, was sentenced in 2000 to 450 hours of community service, mandatory weekly therapy, and was fined $180,500. Compare those sentences to the case of Chrissy Taylor, a thirty-year-old California woman who has already served ten years of a two-decade prison sentence imposed for purchasing legal chemicals that her boyfriend used to make drugs, or to the plight of Clarence Aaron, a nonviolent first-time offender who is now serving life without parole in an Atlanta prison for introducing two drug dealers to one another. Taylor was an impressionable nineteen-year-old whose main mistake in life was hooking up with the wrong guy; Aaron was a troubled senior at college in New Orleans when he committed his crime.

Or consider the disparities in sentences meted out to those who rob banks. Although the brazen looting of savings and loans in the '80s cost taxpayers nearly $400 billion, very few savings and loan crooks were given prison time. Even the most infamous of the savings and loan cheaters, Charles Keating of Lincoln Savings and Loan Association, managed to avoid the full penalty for crimes involving $3.8 billion dollars in losses: four-and-a-half years into Keating's ten-year sentence, a federal court ruled that faulty instructions to the jury had tainted his trial and threw out the guilty verdict. Keating's time served was about the same as the sentence given out recently to a twenty-one-year-old Boston man who made off with $1,000 from a bank. Though he had specifically informed bank tellers that he was unarmed, Coleman Nee's crime earned him a fifty-seven-month sentence.[27]

White-collar criminals have also benefited from the wide disparities in sentencing across the nation—an advantage not granted to drug offenders facing uniform mandatory minimums. States such as Wisconsin, which are home to fewer firms offering financial services and other complex transactions vulnerable to fraud, tend to mandate jail time in as much as 85 percent of the cases. In New Jersey, where prosecutors must make deals in order to keep caseloads at manageable levels, white-collar criminals got prison time in only 26 percent of cases.[28]

The coddling of white-collar criminals extends into their prison terms. For fees ranging up to $50,000, you can hire a "postconviction placement specialist" who'll help you get into the Club Fed of your choice. He'll develop a glowing pre-sentence investigation report that details your contributions to humanity and which may run for dozens of pages packed with superlatives and misty-eyed testimony from priests or scoutmasters you haven't seen in thirty years. Then he'll duke it out with the Bureau of Prisons to get you good prison choices. The options aren't that bad for those in the know. Leafing through the *Federal Prison Guidebook,* an insider's guide to Uncle Sam's facilities, you may be reminded of your college dormitory days. Perhaps your first choice will be the Schuylkill federal prison camp in Pennsylvania, where fallen biotech enterpreneur Samuel Waksal now resides. This modern complex has a crafts center, outdoor sports teams, game room, and much more. Or, if you don't like the cold and do like billiards, you may consider the popular Eglin Federal Prison Camp in Pensacola, Florida, with a top-notch exercise room, softball fields, pool tables, and big-name alums like E. Howard Hunt and Aldo Gucci. Then again, maybe those humid summers are too much for you, in which case you'll be wise to consider the minimum-security camp in Lompoc, California, where Ivan Boesky did his soft time, as did the Watergate offenders.

Lompoc is an hour away from Santa Barbara and has perfect weather year-round. There is a baseball diamond and a volleyball court. The population of inmates at Lompoc has gotten rougher in recent years, but it's still a great place to catch some sun and get in shape, although ultimately, like any of the prison camps, "it's jail, not Yale." No cell phones, no PalmPilots, no toupees, and no day trading from a laptop in your cell. Also, the food stinks.[29]

Whether they go to prison or not, convicted white-collar criminals nearly always face fines and settlement penalties. These are impressive only to the untrained eye. The federal government's dismal record of collecting fines assessed in white-collar crime cases illustrates another way that wealthy felons evade serious consequences. As the number of white-collar prosecutions increased in the late 1990s, the amount of criminal debt demanded for restitution and compensation has grown significantly, from $5.6 billion in 1995 to more than $13 billion in 2002. But the government's unwillingness or inability to collect those debts has rendered the increase largely irrelevant. The Justice Department lacks any systematic method for evaluating forfeitable assets, establishing payment schedules, or monitoring compliance after prison and probation. The result has been a massive flouting of fines.[30]

Edwin McBirney II of Dallas, for example, was fined $7.46 million for his role in looting Sunbelt Savings in the 1980s. As of 2002, according to *US News & World Report,* McBirney had paid only $32,910 of his debt, even as he enjoyed the amenities of a $1 million home and a limousine and driver. Larry Vineyard of Englewood, Colorado—another savings and loan scoundrel—paid $1 million of his $5 million fine in 1996, when prosecutors threatened to bring charges against members of his family. Since that time, he has not met any of his $50 monthly installments. "There just seemed better things to do with the money," explained Vineyard, whose driveway in Dallas is crowded with four late-model SUVs for his personal use. A California man has not paid a cent of the $1.2 million fine imposed for his role in a 1987 golf course scam—even as he has maintained homes in California and Mexico, a yacht, and two private airplanes. Neil Bush, George W. Bush's younger brother, tarnished his family's reputation over a decade ago when he was named in a $200 million suit by the FDIC for his role in the failure of the Silverado Savings and Loan. But the episode didn't compromise the family fortune: Bush settled with the FDIC for only $50,000.[31]

Ivan Boesky had the best laugh of all at the expense of prosecutors. Boesky took advantage of a loophole to claim a substantial tax benefit as a result of his punishment for massive insider trading. By classifying his $100 million fine as a business expense, Boesky was able to deduct $50 million from his federal income tax in 1986. Merrill Lynch scored a similar tax windfall by writing off a third of the $100 million settlement imposed by Spitzer. Also, depending on their tax rates, many corporations that set up funds to compensate the victims of financial crimes can deduct up to 40 percent of these costs.[32]

White-collar criminals also hold on to their assets in other ways. The increasingly common practice of incorporating businesses offshore, as well as moving personal assets into offshore accounts or into complex overseas investment shelters, offers legal protection to savvy wrongdoers. Pensions and personal homes are protected from seizure in some states and have become additional vehicles for shielding wealth from the ever shorter arm of the law. The civil case against O. J. Simpson drew attention to the ways in which personal retirement funds are protected against seizure and can shelter tremendous wealth. Despite losing a multimillion-dollar lawsuit against Nicole Brown's family, Simpson is living well on $25,000 a month in pension income that the civil judgment can't touch. Because Simpson is a resident of Florida, the law is also unable to touch his home. Like Texas, Florida has a generous "homestead exemption" law that prevents creditors or law enforcement agencies from seizing personal homes. The homestead exemption in Texas has made it possible for Enron's Ken Lay to hold on to his 13,000-square-foot, $7.1 million home in Houston. In Florida, the Sotheby's price-fixer Diana Brooks does not have to worry about the seizure of the $4 million waterfront home that she and her husband purchased after the government investigation of Sotheby's was under way. Scott Sullivan's palatial mansion under construction in Boca Raton also probably won't be touchable, however large the fines or settlements that result from his role in WorldCom's $11 billion fraud.[33]

ECONOMISTS ENDLESSLY ARGUE that human actions are shaped by rational calculations about costs and benefits. This generalization doesn't always hold—witness the irrational influences of religion, love, or rage—but it is largely true.

The lax treatment of cheaters at the highest levels of our society inevitably shapes the calculus of anyone who contemplates cutting corners. It is just too easy in this society for cheaters to float seamlessly upward, seeing few downsides along the way. There would be nothing extraordinary today about someone who cheats academically in high school to help himself get into a good college where he continues cheating with few consequences, even if he is caught red-handed. Later, with the help of a résumé that includes fictional elements that go unchecked, he could land a job in business. Every year he could help himself out financially by cheating on his taxes with very little risk that the IRS will audit him and, should the IRS do so, that the government would be able to do anything more than order him to pay back taxes. Meanwhile, at the office, should he enrich himself through a sophisticated scheme to defraud clients or investors, he'll have little fear that his actions will result in either legal or financial penalties.

In an absolute worst-case scenario—if he really steals a lot of money and the evidence against him is overwhelming—the Sarbanes-Oxley Act passed by Congress in 2002 might ensure that he does some time in a federal prison camp. (Then again, maybe not. While the new law has been hailed as tough because it raised the maximum prison sentence for securities fraud to twenty-five years, prosecutors have wide discretion in how they classify crimes and judges have discretion at sentencing.) If he does do time, his friends and family will be appalled, and he'll be mortified about his downfall. But, according to research on white-collar criminals, he's likely to bounce back quickly once he gets out of prison.[34] He'll emerge from Club Fed twenty pounds lighter and in the best shape of his life. Thanks to prison rules, he'll have lost the silly toupee that he used to wear back in his days as a wheeler and dealer. If he had half a brain when he was making money, there'll be plenty still stashed away to help him adjust to his life as an ex-con. He'll have a large nest egg in a bank in Grand Cayman, or some such place, an invulnerable pension, and maybe a house in Florida. Assets wisely transferred earlier to his wife and kids will be untouched. There will be an outstanding fine or settlement against him, but he needn't worry about that. He can focus on getting on with the quintessential American job of creating a second act—maybe one even loftier than the first.

The first few years will be hard, especially those days when he must report to his probation officer. But eventually his crimes will be forgotten, or explained away as a consequence of heavy-handed government regulators run amuck—an excuse that will find much sympathy at the club where he plays golf. Pretty soon he'll be back on his feet with a new career. Generous donations to the right charities will help lubricate his reentry into respectable society and, in time, he will no longer be shadowed by his past.

Michael Milken stands as the most encouraging role model. Milken was sentenced to ten years for his role in the massive frauds on Wall Street in the '80s, but served less than two years. Despite a huge government fine he emerged from prison with hundreds of millions of dollars in family assets. A wave of "Milken revisionism" emerged in influential quarters during the 1990s. Even Milken's former critics proved susceptible to the financier's multifaceted public relations campaign, which featured Milken's role as a cancer survivor and activist, his extensive philanthropy, and the achievements of his California-based think tank, the Milken Institute. (By 2000, the work of the Milken Institute was funded mainly by the founder's annual contribution of $5 million.) Though the terms of his sentence expressly prohibited his participation in financial enterprises, Milken earned nearly $100 million in consulting fees for financial transactions between 1993 and

1996. This work included television acquisitions for Rupert Murdoch's News Corporation and the 1996 purchase of Turner Broadcasting Service by Time Warner, Inc.[35]

As Bill Clinton prepared to leave office, the *Wall Street Journal* and other mainstream publications called on the president to pardon Milken and clear his name. Presidential pardons represent the mountaintop of rehabilitation these days. Yet for those pardon seekers who have no redeeming characteristics, this mountaintop can only be reached via mammoth lobbying expeditions financed by campaign contributions. Milken evidently didn't make the investment. Denise Rich did make that investment, pledging $450,000 to the Clinton library, and her ex-husband, Marc Rich—a brazen swindler in his day—got the magic signature from Clinton. (Even before Rich was pardoned, his Switzerland-based commodities firm had won $65 million in U.S. grain-export subsidies as well as a $20 million contract to sell metals to the U.S. mint. Rich's coziness with the U.S. government, it should be said, was not unusual. The federal government has a long history of awarding contracts to corrupt companies and entrepreneurs who have repeatedly broken the law and been heavily fined. Getting on Uncle Sam's bad side is yet another negative that cheating business leaders needn't worry about.)[36]

Milken's afterlife has been more lustrous than that of his former partners in crime, but they have done well, too. Ivan Boesky emerged from prison as a fitness nut, with a deep tan and bulging biceps. Boesky's personal secretary, among others, reported that the financier enjoyed lavish accommodations and decadent amusements in the months after his release from Lompoc Federal Prison Camp. Visiting his mountaintop retreat, said Janice Rheel, the secretary, "makes me feel dirty. There's stuff going on that I just don't feel comfortable with. It's the young girls, other stuff I can't talk about." While much of his personal fortune was lost to government fines and restitution charges, Boesky retains his grip on the good life by virtue of a divorce settlement that required his wife to pay $20 million in cash and $180,000 a year in alimony. Boesky also was granted ownership of the couple's $2.5 million home in La Jolla, California.[37]

Dennis Levine, the banker whose initial arrest led prosecutors to Boesky and finally to Milken, is also doing well. He resumed his career in finance after his release from prison in 1991. As the president of Adasar Group, Inc., Levine earned enough to afford an elaborately appointed 2,200-square-foot Manhattan apartment.[38]

By far the most successful of Milken's former pals is Gary Winnick, who worked closely with the junk bond king at Drexel. Winnick narrowly escaped prosecution for his involvement in Milken's crimes by agreeing to testify against his former boss. Winnick was never called to the witness stand and he went on to have a glorious new life on the shady side of the telecom industry. By 2000, Winnick was chairman of Global Crossing, a company worth over $50 billion, at least on paper. Winnick cashed in over $700 million worth of Global Crossing stock before the company went bankrupt in 2002. He used his new fortune to buy, among other things, a private jet and a California estate valued at over $90 million. Although Winnick has been accused of a wide variety of misdeeds, the federal government chose not to pursue a criminal case against him. Winnick is only in his mid-fifties and has plenty of time for a third act. Generous charity donations through the Winnick Family Foundation are sure to help people forget all about Global Crossing.

Other culprits in the big scandals of the '80s and early '90s are also prospering. Bill Walters, notorious for his role in the Silverado Savings and Loan collapse, lives in high style. Walters managed to avoid paying $280 million to his creditors (including a $106 million fine assessed by the government) by declaring bankruptcy in 1993.

Thanks to complex arrangements that had transferred millions of dollars in assets to his wife, he retains a good chunk of his ill-gotten fortune.[39] Walters's old associate from the Silverado days, Neil Bush, rebounded even more admirably from his brush with the law and financial ruin in the late 1980s. By 2002, Bush was CEO of an education software company that had raised more than $20 million in investment capital. He and his wife appeared regularly in Houston papers as the hosts of high-profile charity events, while his daughter Lauren Bush, an up-and-coming fashion model, engaged in a publicized romance with Great Britain's Prince William. And, of course, there were always visits to the White House—which his brother won in 2000 in an election characterized by far-reaching chicanery among partisan Florida election officials.[40]

High-profile cheaters outside the business world also do well after their fall. Sandra Baldwin, disgraced by a faked résumé, returned to the real estate business in her hometown of Phoenix after her resignation from the U.S. Olympic Committee. Her contacts and drive enabled her to quickly become the top company producer for Coldwell Banker Success Realty. George O'Leary, another person with an invented past, was also nearly instantly back on his feet after being forced out as Notre Dame coach. O'Leary now has a lucrative job as an assistant coach for the Minnesota Vikings.[41]

When it was recently revealed that disgraced *New York Times* reporter Jayson Blair had landed a book deal worth a mid-six figure, many people were dismayed that Blair was profiting from his lies. But then Blair was simply following the lead of Stephen Glass, the *New Republic* writer who resigned in 1998 after admitting that he concocted some of his stories. Glass landed a six-figure book deal from Simon & Schuster, and the publisher mounted a major push to promote *The Fabulist*, Glass's fictional account of his odyssey, which provided a sympathetic self-portrait of an ambitious young man who makes some understandable mistakes. Glass was featured on *60 Minutes*, in the *New York Times*, and other venues that most authors can only dream of. Given the nature of American culture, Glass's ability to profit handsomely from his bad behavior was hardly surprising; perhaps more surprising was that Glass was actually admitted to Georgetown law school after his journalism career dissolved. A history of lying, the admissions office apparently concluded, prepared Glass perfectly for a career in law.

Even Tonya Harding has done all right for herself. While the men who smashed Nancy Kerrigan's knee received prison sentences, Tonya Harding got off with a slap on the wrist. Her sentence included $160,000 in fines and 500 hours of community service. This modest downside was balanced by a considerable upside. Harding received a reported $600,000 in exchange for her confession on *Inside Edition*.

One year after the attack on Kerrigan, Harding was already enjoying a rehabilitation in her public image. Both *People* and *Esquire* placed her on their annual lists of favorite celebrities for 1995 and respondents in one survey listed Tonya Harding as among the twenty athletes they admired the most. Harding had fine company on the list: alleged murderer O. J. Simpson and convicted rapist Mike Tyson. Harding went on to launch a new career not only as a professional boxer, but also as a "celebrity boxer." In March 2002, Fox Television paid Harding $50,000 to duke it out on national television with Clinton-scandal star Paula Jones. (Harding trounced her.) The show attracted 15.5 million viewers and was one of the highest-rated shows in Fox's season. Recent years have also seen Harding invited by ESPN to a skating contest, featured on *Entertainment Tonight*, given an acting part in an HBO comedy, invited to host programs on TNN during "Bad Boys Week," asked to appear on the TV game show *The Weakest Link,* interviewed twice by Larry King on CNN, and—of course—signed up to write her autobiography.[42]

That's the way it often goes in an America that loves second acts. Fame and money lead to more of the same—while past sins are conveniently forgotten.

CHAPTER NINE

Dodging Brazil

NEARLY EVERY YEAR FOR THE PAST DECADE A GROUP OF M.B.A. students from Pepperdine University in California have made an unusual road trip to Las Vegas. They go not to gamble or catch glitzy shows. They go, ironically, to polish their conscience.

The visit kicks off in a hotel conference room with evening lectures by two former business executives, Mark Morze and Ted Wolfram, who have spent time in prison for white-collar crimes.

Mark Morze is the more dynamic of the two men. A loud, high-energy man, he used to be the chief financial officer of ZZZZ Best Carpets, where he helped orchestrate a $100 million fraud. ZZZZ Best claimed to be making millions of dollars cleaning carpets when, in fact, its sole function was to rip off investors. Morze did seven-and-a-half years in prison. Ted Wolfram is an older man, a devout Catholic who used to be a banker in the Midwest and served ten years after he started making loans to himself—to the tune of $47 million. The clear-cut thefts of both men set them apart from many white-collar criminals in that they actually got serious prison time.

The Pepperdine students are mostly midlevel business managers on leave from their jobs or working nights and weekends to get through Pepperdine's executive M.B.A. program. They listen, rapt, as Morze and Wolfram explain the missteps that led them into a life of crime and to prison. The two ex-cons filter their past in very different ways. Wolfram is a sad man, who talks of forever losing his honor. "I spent ten years in prison and deserved every minute of it," Wolfram likes to say. Mark Morze seems to still be the pushy con artist he was in his ZZZZ Best days. Morze does these speeches for a living and, with the recent corporate scandals, he is very much in demand, taking home between $3,000 and $5,000 a speech. He tells funny stories about fabricating documents or sharing a prison cell with a former congressman. "The students have commented that they listen to Mark and realize that he's sorry he got caught, and listen to Ted and realize he is sorry for what he did and the effect it had on his bank and his company," says Scott Sherman, a Pepperdine professor who helps organize the sessions.

The morning after the presentations in Las Vegas, the Pepperdine students journey by bus to Nellis Prison Camp, a federal minimum-security prison located on the Nellis Air Force Base outside of Las Vegas, which houses a number of white-collar criminals. The prison visits were started in 1987 by Pepperdine finance professor Jim Martinoff, and they are intended to give M.B.A. students a hard look at the consequences of corporate crime. The program is reminiscent of the Scared Straight initiatives begun in the '80s which brought groups of young juvenile delinquents to state penitentiaries to meet hardened convicts and get a glimpse of what lies in their future if they don't change their ways.

"Usually there is a panel of about six inmates," explains Sherman. "Each is given forty-five minutes to talk about what they did and what message they want to get across to the students, and what they would do differently." The longer the inmates have been incarcerated, the more honest they tend to be about how they screwed up. "Generally speaking, they were successful," Jim Martinoff says about the inmates. "They were in a rat race and had to get better and better. Their crimes start small in a gray area, and then they get into a little more of a gray area, and then they do more.... One of the things they always say is 'Everybody else was doing it and I needed to be competitive, otherwise the company was going to fail.'"

The vivid tales of how each of the inmates sprawled down a slippery slope often have a profound effect on the M.B.A. students. "Some students get uncomfortable," says Sherman. "Something is said that reminds them of something going on at work, or something they are thinking about doing—they move near the back until they are practically hugging the back wall of the room.... Usually the ones in the back of the room are trying to disengage and don't ask questions. One of the classic questions was an executive from Xerox, a wonderful woman, who asked, 'At what point in time did your gut tell you that maybe you were doing something you shouldn't have been?' The inmate said, 'It was once I was no longer willing to share with my spouse and my family what I was doing.'"

Sherman says about this exchange: "That has become what we refer to as 'the mama rule.' If you can't tell mama, why are you doing it?"

Sherman and Martinoff also talk of a "2 percent rule." That is, about 2 percent of the students who go through the prison visit are deeply unsettled and change their career as a response. Some of these students also tell Sherman and Martinoff that they need to contact law enforcement officials to discuss things happening at their company.

The business school at the University of Maryland also has a Scared Straight component in its curriculum. Visits to the federal minimum-security prison in Allenwood, Pennsylvania, are scheduled for the weeks immediately prior to graduation, in hopes that memories will remain fresh as their graduates face real-life ethical dilemmas in the workplace. This initiative, too, seems to have an impact on students. "We're brought up on the concept of competition," explained one M.B.A. student after the prison visit. "But once you get into that competitive spirit and the adrenaline gets going it's easy to take shortcuts and overlook principles." Having witnessed the high price of noncompliance, the student said he would "take extra precautions to be sure I'm on the right side."

There is no data on the long-term effectiveness of dragging business-school students into white-collar prisons, and only a few schools undertake such visits. Maybe this has an impact, maybe not. But these Scared Straight initiatives are impressively energetic and imaginative efforts to prevent cheating. More of this same spirit is needed nowadays, and then some.[1]

America's epidemic of cheating does not lend itself to easy antidotes. While cheating does boil down to choices made by individuals, these choices are heavily shaped by cultural, political, and economic forces. Even people determined to lead a life of integrity may find themselves cheating at different moments—and rationalizing it with surprising ease.

Taking on America's cheating culture requires taking on the societal forces that are driving this epidemic. It also entails new crusades for integrity in our schools and workplaces. I see the work ahead as broadly boiling down to three tasks: We must forge a new social contract in the United States. We must reform key professions and instill new codes of conduct in the workplace. And, to have any hope of an honest future, we must strengthen the ethics of new generations of Americans.

Sounds like a piece of cake, right?

A New Social Contract

Cheating thrives where unfairness reigns, along with economic anxiety. It thrives where government is the weak captive of wealthy interests and lacks the will to do justice impartially. It thrives where money and success are king, and winners are fawned over whatever their daily abuses of power. If we want to reduce cheating, we must tackle these root causes. We need to create a new social contract in America that gives people faith in a few simple principles: Anyone who plays by the rules can get ahead. Everyone has some say in how the rules get made. Everyone who breaks the rules suffers the same penalties. And all of us are in the same boat, living in the same "moral community" and striving together to build a society that confers respect on people based on a wide variety of accomplishments.

Rediscovering these principles doesn't require an archaeological expedition. These principles are at the heart of the American Dream; they are deeply embedded in our psyche. Nearly all Americans embrace them at some level. This ethos is what the United States, at its best, is all about.

But as has happened before, these basic tenets of Americanism are being elbowed aside by another central element of our national identity, namely, unfettered capitalism. We are learning once more—as leaders like Teddy Roosevelt learned a century ago—that our noblest aspirations for America cannot be realized when we allow economic competition to grow too harsh. Creating a new social contract requires taming the free market, as we have done in past eras. It means striking a healthier balance between humanistic values like shared responsibility, mutual respect, and compassion for the weak—and market values like maximum efficiency, individual autonomy, and admiration for the strong. Both of these sets of values have deep roots in our society. In fact, you can think of most Americans as having a split personality, as my colleague, Tamara Draut, has put it. The Tough American embraces the rigors of the market as the best way to ensure wealth creation and personal liberty. She isn't interested in a lot of coddling of those who reject hard work. But there's another side to her, the Fair American. She wants to live in a society where everyone's treated equally and the weak aren't left by the roadside. Creating a new social contract means finding the right public-policy medication to deal with America's split personality. These days we need less Tough American and more Fair American.

Nowhere is this more true than in the economic sphere. Competition has grown too harsh in our postindustrial Banana Republic economy. Dog-eat-dog norms have conspired with long-term trends to create, on the one hand, fantastic concentrations of wealth and, on the other, profound insecurities among a vast swath of Americans. Neither is conducive to a society with strong ethics.

While crusades against inequality typically don't get very far in the United States—Americans admire the rich too much to want to redistribute their wealth to the masses—what can fly are efforts to help more Americans join the middle class, or feel more financially secure as members of the middle class. And today, there are plenty of commonsense ways to work toward these goals, ways that appeal to both sides of our split personality.

First, we need to make work pay. One of the cardinal principles of the social contract in America is that everyone who can work should work, and that those who work have a chance to climb up the economic ladder. Yet today at least a third of working households in the United States can't even make ends meet, much less secure a place in the middle class. Raising the minimum wage to around $8.50 an hour and then indexing it to inflation is one way to make sure that people who work are decently rewarded. Polls show very strong public support for a higher minimum wage—over 80 percent in most surveys—and research shows that the impact on business would probably be minimal, despite the hysterical warnings of industry groups. Tax credits offer another promising strategy to help low-income families get their heads above water. The Earned Income Tax Credit now funnels over $30 billion a year in tax refunds to families making under $30,000 a year, and it has historically enjoyed strong bipartisan support. Making the EITC more generous and complementing it with other tax credits targeted at the bottom third of households would also go a long way toward guaranteeing the basic promise of work, namely, that it gets one ahead in life.[2]

Likewise—and perhaps most critically—high levels of economic growth are needed to ensure a payoff from work and an expanding middle class. When the economy is booming, tight labor markets help ensure that wages go up for everyone, even for workers at the bottom. When the economy slows and unemployment rises, wages go down. How to keep things booming? Among other things, the federal government must be ready to prime the pump in ways that yield long-term dividends—through investments in transportation infrastructure, scientific research, energy efficiency, and so on. The best government intervention in the economy kills two birds with one stone: it keeps growth zipping along while laying the groundwork for even greater future prosperity.[3]

A second strategy to help people get ahead is to expand access to higher education. Want to make something of yourself? Then go to school and work your ass off. So says the Tough American, and the Fair American couldn't agree more. Yet lately, even as the United States has moved into an information age in which education counts more than ever, we've been cutting support for higher education. Students are paying higher tuition and shouldering heavier debts when they graduate. While more kids than ever are starting college, many can't afford to finish and others can't go at all. Here again, people who are playing by the rules confront a system that is stacked against them. Legions of undereducated workers languish in dead-end service jobs—not as a stepping-stone to something better, but as a permanent reality. And guess what? They aren't too happy about this.

A major new investment in higher education would help restore a critical element of America's distinctive social contract—that anyone who wants to improve their station in life can use education to achieve this goal. The United States should seriously commit itself to building a system of universal higher education for the twenty-first century. Most of the components of that system are already in place, including ubiquitous community colleges across the country and strong state universities. What's missing is the money. A new initiative that marshals the resources of public, private, and nonprofit sectors is needed to ensure that every young person who wants a K–16 education—or serious vocational training—can get one through a combination of affordable tuition, grants and scholarships, and low-interest loans. Easier said than done, I know. But this is an essential foundation for fairness in the new century.

Third, we need to help people build wealth and personal assets. Wealth has traditionally been a linchpin of the American Dream. If you want to build a better life for yourself and your children, the Tough American will tell you to tighten your belt and think long term: pay off a mortgage through years of hard work; skip the new

DVD player and build your savings; work fourteen-hour days at a family business that can be passed down to the next generation. The Fair American agrees with all of this. What's lacking in the United States right now, though, is the opportunity for many people to build wealth. Too many families aren't making enough money to save for the future and instead are accumulating high-interest debts. Most people don't have financial reserves they can fall back on if they lose their job or have a health crisis. Homeownership rates may be at a historic high, but a third of families still aren't part of the ownership class, and many people who are ostensibly part of this class actually have a negative net worth. People without any savings, or those carrying a lot of debt, move through life shadowed by insecurity and may find it easier to rationalize doing "whatever it takes" to improve their financial situation.

Smart proposals abound for how to help more people create personal wealth: birth endowments that give every child a nest egg on day one and, through the miracle of compound interest, translate into real assets by adulthood; Individual Development Accounts that leverage government money to encourage poorer Americans to save; special housing programs that offer low-interest loans to first-time homeowners; micro-credit loans that allow more people to start a small business. The ideas are all there. What is needed now are serious investments aimed at creating a true "stakeholder society."[4]

Fourth, more needs to be done to reduce key insecurities that are part of our postindustrial economy. Americans may never again have the kind of job security that was common forty years ago. The days of strong labor unions and benevolent employers who provide good benefits may never return. More and more workers in the new economy will increasingly operate as free agents and have responsibility for meeting their own pension, health-care, and child-care needs. All of this can be a good thing, giving people more freedom, organizations more flexibility, and the economy greater dynamism.

Yet even the Tough American will agree that workers today face too many insecurities. As many as 60 million Americans are without health-care coverage at some point during any given year. Nearly half of workers are not covered by employer pensions and are not saving enough to retire securely. Tens of millions of working parents face an ongoing child-care emergency that limits their ability to participate in the economy and causes intense stress. Many workers don't have enough flexibility in their jobs to deal with a family illness and other personal problems—or the resources to sustain themselves financially when they are out of work for very long.

None of the Tough American's nostrums are going to get America out of this pickle. More hustle or more planning for the future is not going to solve these insecurities, which are caused by a basic failure of the market to deliver vital public goods. Instead, only the government has the resources and clout to help ensure affordable health care, child care, and housing. A new commitment along these lines—in partnership with business and nonprofit groups—is long overdue.

Taken together, these four strategies would reduce the financial anxieties that stalk many households and help restore people's faith that the "system works," that if they play by the rules they will be rewarded with a decent life. These steps should appeal to both sides of the American character: each asks a lot of people, as the Tough American demands; yet each holds out real rewards, as the Fair American suggests. And all four envision government not as the sole solution to our problems but as a catalyst for action that leverages its resources in partnership with the private and nonprofit sectors.

How should we pay for all of this? Alas, that is always the big sticking point. As Jennifer Hochschild and other scholars of the

American Dream have discussed, it's one thing for people in principle to embrace core American ideas of equal opportunity; it's another thing for them to put up the money so the nation can realize these ideals. This is particularly true around education, where Americans at once deplore—and tolerate—"savage inequalities" in opportunity.

At the same time, if Americans can reach a deeper agreement on why and how to foster equity, the obstacle of money might not loom as large. If we find analyses and approaches that unite us ideologically, we'll have fewer things to fight about and it will be easier to muster the resources needed for the job. Clearly the money exists in our $10 trillion economy. In his book, *The Two Percent Solution,* Matthew Miller lays out a far-reaching bipartisan strategy for fulfilling the promise of equal opportunity—one that parallels some of the ideas above and would cost just 2 percent more of the national wealth we generate annually to implement.

A word of caution, however: expanding opportunity will not get rid of inequality. Even if we heavily tax the rich, high levels of inequality are here to stay thanks to changes in the economy. Public policy can't alter the fact that someone with a Ph.D. in biology or a natural knack for writing software code is going to be paid a lot more than someone with a certificate in dental hygiene or a gift for nursing the elderly. The winner-take-all dynamics of economic competition are especially intractable. Government can't realistically do much about a ball team that wants to pay sluggers fifty times more than benchwarmers or a university that wants to pay star professors ten times more than adjuncts. There will always be a Winning Class, along with legions of people desperate to join it. And there will always be those who are winners and yet feel that they don't have enough money or status. In the end, we will never fully eliminate the gap between what people want and what most of us can get through legitimate means. Even if we did greatly reduce inequality and achieve broad gains in absolute well-being across U.S. society, many people would still think negatively about their relative position. Wherever they are on the economic ladder, people will always be able to gaze longingly upward at someone who has more. Incentives to cheat will endure.

This reality means that it is all the more important to pursue equality in the political, legal, and social arenas of American life. To curb the cynicism among ordinary Americans and to rein in the hubris of wealthy Americans, the laws governing our society must be made fairly, enforced fairly, and seen as fair.

For starters, we must adjust the scales of influence within our political system. You can forget about getting money out of politics entirely, mainly because of the First Amendment. But money will talk less loudly if we have public financing of elections and also provide free television and radio airtime for candidates. Both of these measures would allow candidates to run viable campaigns—without prostituting themselves to wealthy donors. Curbing lobbying through stricter rules on the gifts politicians can accept and what jobs former government officials may take is another reform that is needed to rein in the influence of the wealthier classes and corporations.[5]

Meanwhile, ordinary Americans need to get more involved in elections and thus in shaping the laws that govern their lives. Instant runoff voting that allows voters to rank candidates in order of their preference would give the public more choices and more reasons to show up at the polls. Allowing voters to register on election day (a process that exists in six states) would also make a difference, since most people only get interested in electoral contests—if they pay attention at all—in the final few weeks or even days leading up to an election. Making election day a holiday, or holding elections on a weekend as other countries do, would mean that people don't have to squeeze voting into their workday.[6]

Cracking down on the financial cheating of wealthier Americans and corporations must be the next order of business, although it won't be easy. Three steps are particularly important to reverse the trends of the past two decades: give the IRS real muscle to ensure fairness in tax collection efforts; strengthen the SEC and better police financial markets; and give law enforcement agencies more resources to put away white-collar crooks.

The fairer enforcement of tax laws is an obvious priority. Not only should the IRS get the huge budget increases it needs to do its job, but it should also get new marching orders that compel equity in how it pursues tax evasion. Enforcement priorities should be determined by whose cheating is most costly, as opposed to who the easiest targets are. Following the money will naturally lead to a much greater focus on upper-income earners, a shift which is long overdue. But it is important that these earners do not feel unfairly singled out by the IRS. In the end, our tax laws will only have legitimacy if they are applied equally—and if all Americans perceive that they are applied equally.

The financial markets are another key focus for any effort to scrub clean the upper reaches of American society. The reforms enacted in 2002 and 2003 mark an important step forward in policing financial markets and corporate behavior. The Sarbanes-Oxley Act makes CEOs legally accountable for the numbers contained in company earnings reports and dramatically increases the penalties for financial fraud—up to twenty-five years in prison in some cases. It also mandates that corporate directors be more independent, and it creates a new public office to oversee the accounting profession. Elsewhere, the 2003 agreement between leading investment banks and regulators—led by Eliot Spitzer—calls for big changes at financial services firms to reduce lying by research analysts and other abuses. Finally, new legal and regulatory changes promise to help clean up accounting firms by prohibiting these firms from doing consulting work for the same companies that they are auditing.

In short, we do seem to have learned something from the recent corporate scandals. Yet while all these reforms are important, and all are overdue, government responses to the corporate scandals have not gone nearly far enough. The overwhelming majority of high-level wrongdoers have walked away from these scandals unscathed and phenomenally wealthy. As of this writing, the two men at the very top of Enron, Kenneth Lay and Jeffrey Skilling, have faced no criminal action; Bernard Ebbers of WorldCom has not been indicted; prosecutors have declined to go after Gary Winnick of Global Crossing; Jack Grubman and Henry Blodget escaped criminal prosecution in their settlement with Spitzer; Sanford Weill of Citigroup has not been pursued by prosecutors; Philip Anschutz of Qwest agreed only to a small settlement with Spitzer; and the list goes on and on and on.

There are good reasons why these villains have not been punished—mainly that successful legal action against them is difficult given the complex or ambiguous nature of their misdeeds. But the message being sent here is highly troubling. Future wrongdoers may not be deterred; and the public, for sure, will come away from this period feeling even more cynical about the notion of equal justice.

Worst of all, government still lacks the capacity—or legal authority—to fully police financial markets. "The truth is that the markets have been, and are, spinning out of control," writes regulatory expert Frank Partnoy. "The antiquated system of financial regulation, developed in the 1930s and designed to prevent another market crash after 1929, no longer fits modern markets... the markets are now Swiss cheese, with the holes—the unregulated places—getting bigger every year, as parties transacting around legal rules eat away at the regulatory system from within." Partnoy observes that the value of trading in volatile financial derivatives, where people bet on the future value of anything from the yen to mortgages, is now larger than the trading in stocks and bonds. He and many other observers fear that it is only a matter of time before another wave of scandals sweeps through the financial markets.[7]

In July 2002, Alan Greenspan testified before a Senate panel that "It is not that humans have become any more greedy than in generations past. It is that the avenues to express greed have grown so enormously."[8] The new laws designed to curb yesterday's abuses, along with a little extra money for the SEC, fall way short of what is needed to block off these avenues. The federal government needs the regulatory muscle to do more than just keep its head above water; it must be able to keep pace with more sophisticated attempts to manipulate financial markets in order to anticipate where future abuses will occur and to take regulatory action before those abuses unfold on a large scale. One positive step would be to turn the SEC into a far larger and more entrepreneurial agency—beginning by allowing the agency to "eat what it kills."

The SEC currently takes in up to $2 billion annually in processing fees for corporate filings. It rakes in other funds from fines and settlements. Congress originally intended such funds to go directly toward covering the cost of SEC operations, but they now go into general government revenues. Allowing the SEC to keep all the funds it generates would enable it to grow the fangs it needs and stay on top of sophisticated global financial markets that operate twenty-four hours a day at lightning speed. It could pay much higher salaries to rope in the talented professionals it needs to grasp the regulatory challenges posed by high finance in the twenty-first century; and it could staff an enforcement division that has the capacity to unravel sophisticated scams, as well as to go after all the wrongdoers that catch its eye—as opposed to always operating in triage fashion due to budgetary constraints.

The effective policing of financial markets also requires a major new investment in prosecuting white-collar crime. The SEC has no authority to throw people in prison; it can only bring civil complaints and refer cases to law enforcement agencies. Justice doesn't get done if these law enforcement agencies lack the resources to go up against individuals who rank among the world's most sophisticated and wealthy criminals. Deterring and punishing financial crimes—but also white-collar abuses in other areas—requires a major expansion of federal prosecutorial powers. That expansion should begin by dramatically beefing up the securities fraud unit in the U.S. attorney's office in New York, which will remain ground zero in the fight for clean financial markets. Yet in this day and age of the populist stock market, with more brokerage services operating in every part of the United States and large public companies spread across the country, law enforcement expertise around securities fraud must also be beefed up in many other places beyond New York.

On another front, law enforcement agencies must step up their assault on illegal practices within the health-care and pharmaceutical industries. These industries have become among the most corrupt in the United States. At a time when health-care and prescription drug costs are already spiraling out of control, and the coming old age of the baby boomers promises to badly strain health-care entitlement programs, various forms of fraud are costing the nation tens of billions of dollars a year. This trend is unfair and demoralizing. Even as many Americans are unable to afford decent health care, they hear constantly about stealing and profiteering by wealthy pharmaceutical companies and hospital chains. The repeated failure to punish specific individuals responsible for these crimes can only deepen public anger around a health-care system that already stands as the leading symbol of a broken social contract.

Law enforcement agencies are already getting serious about health-care rackets. The Justice Department has recently issued new rules governing the marketing of prescription drugs, and it has also gone up against large health-care providers in several high-profile cases over the past few years. Typically, though, the cases against wrongdoers have taken years to fully investigate and few tough penalties have been meted out. For example, the government's case against Parke-Davis for its illegal marketing of Neurontin has already been under way for years and will probably continue for several more. The Columbia/HCA case of the 1990s involved massive billing fraud—outright theft, effectively—and yet only a few indictments were handed down. Other, smaller cases in the health-care field have followed the same pattern: long, drawn-out investigations followed by a slap on the wrist. The difficulty investigators face in achieving more decisive results is due to the same challenges of prosecuting securities fraud: a combination of ambiguous evidence and a lack of prosecutorial manpower. It's time for some serious muscle in this area.

THE STEPS I've suggested so far to create a new social contract should not be very controversial. Helping more people to get ahead is as American as apple pie. So, too, is the notion of making everyone—rich and poor alike—equal before the law. If you believe, as I do, that cheating partly reflects how far America's culture has drifted from the core values of fairness and opportunity, then it follows that new efforts to align our society with those values should reduce cheating.

But this is only part of what needs to be done. Other ailments must also be attacked, especially pervasive materialism, extreme individualism, and the social alienation that thrives amid weak communities and broken families. These conditions degrade our collective sense of moral purpose—the sense that we are part of something larger and that fellow citizens are teammates in this quest. Absent such shared moral purpose, it seems rational to look after yourself and forget about everyone else.

Changing the social and cultural fabric of our society will be no easy thing. A focus on personal freedom and material well-being is deeply ingrained in our national culture. There is nothing wrong with these traits—except when they grow too powerful, rationalizing our worst impulses and dividing people from each other.

Robert Putnam is one of many thinkers who has suggested ways to strengthen civic bonds in America. Others have put forth proposals for shoring up the family in an era where divorce is common and both parents typically work, and for expanding the role of religion and spirituality in people's lives.[9] I would add four other priorities to the mix.

First, we need to give Americans more opportunities to derive meaning from life—and enjoy status in the eyes of others—that aren't about making money and flaunting it. Historically, such opportunities have existed mainly in spaces that stand apart from the free market, including public service, charity, higher education, and the arts. Such opportunities also existed in parts of the private sector that were somewhat insulated from market pressures, like book publishing, journalism, and long-term scientific research for major companies such as IBM or Xerox.

The bad news is that the places where you used to be able to make your mark without worrying about meeting your quota have been squeezed hard in recent years. The public sector has been demeaned and downsized, while the private sector has become a less tenable place for those who lack commercial sensibilities. The good news is that the nonprofit sector is now bigger than ever, offering people more opportunities to make a living, and a difference, in relative isolation from market values. Some $200 billion a year in contributions from individuals and philanthropic foundations is now flowing into more than a million nonprofit organizations, including private universities, religious institutions, international aid organizations, museums, charities, and myriad community groups. The recent philanthropic downturn aside, the nonprofit sector is expected to expand considerably in the years ahead as the endowments of foundations grow through investment in the stock market and as vast new fortunes are harnessed for charitable ends. Among other things, this means more grants for artists, poets, and documentary filmmakers, and more jobs for pastors, environmentalists, and community organizers.

Looking ahead over the next few decades, there is much we can do to expand peoples' opportunities for rewarding careers outside the confines of an ever-harsher market economy. Sustaining the growth of the nonprofit sector is one clear priority, and a critical thing here is to preserve the estate tax, now under attack from Republicans in Washington, since bequests make up a large portion of monies going into the sector. Renewing the vitality of the public sector is an even bigger priority. More robust government is a prerequisite to economic fairness and equal justice under the law, but we also need a flourishing public sector so that those who wish to give back to their nation or community will be respected and well-compensated. Public school teachers and administrators stand as exhibit A of how much this is not the case today.

A second way to humanize our society and pull people closer together is to create more livable communities. The pervasive sprawl in the United States has had a devastating effect on social connectedness. Americans are spending more time in their cars and less time with family and friends. In areas where you need to "burn a pint of gas to buy a quart of milk," people have little occasion to get to know their neighbors. A spirited citizens' movement is trying to change these patterns, imagining a society where every child is within a five-minute walk of an ice-cream cone; where many more people can easily use mass transit and are not prisoners of their cars; where streets have sidewalks and are interconnected, as opposed to ending in cul-de-sacs; where homes are closer together and front porches, not two-car garages, face the street; and where more "third places" exist outside of work and family that allow people to mingle easily with others.

It's a nice vision. But the livable-communities movement faces an uphill battle in a society that too often lets residential growth unfold chaotically, along terms dictated by real estate developers. During the 1990s, the federal and state governments finally began to provide support to communities that were working to become more livable. But much more needs to be done. Major new investments should be made to buy and protect green space from development, as well as to reclaim polluted urban areas, or "brownfields," for redevelopment and to incentivize rezoning efforts that would allow for livable communities. Here again, public funds should be used to catalyze action, not to solve every problem.[10]

Third, we must tackle the problem of excessive consumption. The grotesque materialism of U.S. society is not only bad because it makes people look endlessly inward, creating competitive anxieties about status, looks, and class position. It is not only bad because it has created intense spending pressures on people at a time when many households have been running in place economically. Nor is it only bad because of its destructive environmental consequences. Excessive personal consumption is also bad because it drains national resources away from more important spending that could strengthen our communities and elevate our quality of life. While the U.S. economy now generates over $10 trillion in wealth every year, our political leaders constantly complain that the nation cannot afford many of the things that the majority of Americans say are needed to improve their lives—like better-paid teachers and smaller classes, subsidized health care and child care, better transportation systems, more parks and open space, more affordable housing, more help in caring for the elderly, and on and on. One reason we cannot afford these things is because so much of the wealth generated in U.S. society goes straight into personal consumption. Nearly everyone is buying more stuff than they need—even as nearly everyone complains that our society lacks so many things it requires.

Public policy over the past half century has aimed to promote consumerism. Now it's time to move in the opposite direction. We should use public policy—and particularly the tax system—to reduce excessive consumption. A progressive consumption tax would provide incentives against consumption, increase rates of personal savings, and generate new government revenues. As the economist Robert Frank and others have described it, such a tax could be structured so as to only affect luxury spending. Consumption up to a certain level, that which is needed to cover all of one's necessities, would not be taxed, nor would money put into savings. Consumption beyond these expenditures would be taxed at a progressive rate, with the rate getting higher the more one engaged in luxury consumption. According to Frank's plan, the consumption tax would be collected through the current system of year-end tax filing, and not through sales taxes.[11]

A consumption tax would address the most intractable aspect of today's consumerism, namely that excessive consumption is often rational. The "right" wardrobe or car or home sends critical status signals to other people and can shape one's prospects for a better life. The only way to counterbalance these self-interested motives for consumption is with a more powerful set of financial incentives that run in a counter direction.

A fourth critical change needed to foster a greater sense of shared moral purpose in the United States concerns racial diversity. Nowhere is America's social contract more frayed than when it comes to race. Many nonwhite Americans feel deprived of opportunities to get ahead economically and locked out of any meaningful role in democratic deliberation. The evidence bears out their perception that the social contract isn't working: nonwhites are less likely to have access to good schools, more likely to have been born into poor families without significant wealth assets, more likely to be trapped in dead-end low-wage jobs that do not include health and pension benefits, and more likely than whites to receive harsh punishments in the criminal justice system. Communities of color also have lower voting turnout, make fewer campaign contributions, and overall have weaker political representation than white communities. Not surprisingly, blacks and Latinos tend to be less trusting of government, corporations, and the media, as well as less trusting of other people.

At the same time, many whites see some minority Americans as not doing their part to uphold the social contract. The powerful white backlash following the Civil Rights movement and the war on poverty was partly rooted in plain racism; but it was also rooted in concerns about higher crime in communities of color and a perception of a lax work ethic, low levels of personal responsibility, and little commitment to self-improvement. Carol Swain's book *The New White Nationalism in America* provides a troubling look at how these sentiments are playing out today in different ways than in the past.[12] While there is much evidence that the racial polarization of American society is ebbing, vast divisions persist. These could worsen as the United States becomes radically more diverse in the next few decades, moving toward a point sometime in the middle of this century where whites will no longer be a majority. There is the real potential for growing ethnic tensions in the United States akin to what has occurred in multiethnic countries abroad. If America heads down this road, nothing else we do to foster shared moral community will matter. Social distrust will deepen as people place narrow ethnic self-interest above all else.

This dark possibility makes it all the more urgent that we work to ensure that every American can become a true stakeholder in our society, with special attention given to communities of color where feelings of alienation and disenfranchisement are most intense. Serious wealth-building efforts are needed to address a racial wealth gap that has left an estimated 60 percent of African American households with no assets or a negative net worth. Communities of color—historically marginalized in the political process—should also be the locus of new initiatives to foster civic education and participation. Jurisdictional boundaries that isolate nonwhite urban communities from surrounding suburban areas must be redrawn. And affirmative action must continue until the day comes when leaders across all the nation's institutions have legitimacy in an America of tomorrow destined to look very different than that of today.

A Different Bottom Line

Far-reaching efforts to create a new social contract in America are only part of the solution to today's cheating epidemic. To crack the culture of cheating, we must reform business and leading professions through a combination of government pressure from the outside and change from the inside. Along the way, the private sector must unshackle itself from narrow, bottom-line ideas that too often foster dishonesty.

In 1987, William Morrow & Company published *The Complete Book of Wall Street Ethics*—a gag book of blank pages. The book underscored the view of many that the term "business ethics" was something of an oxymoron. However, while ethics codes may often only scratch the surface of deeply amoral systems, they do offer important ways to attack cultures of dishonesty or prevent such cultures from emerging.

Business ethics got a major boost in 1991, when the Federal Sentencing Guidelines were established. Under the guidelines, a company's penalty for any wrongdoing is determined in light of its commitment to preventing ethical lapses through codes of ethics and compliance programs; strong ethics programs could save companies a lot of money if they ever got into trouble. By 2000, most companies had a written code of ethics and many offered some training in ethics—a big increase from the early 1990s. A 1998 study showed correlation between beefed-up ethics programs and more ethical behavior in U.S. corporations.[13] But other experts have cast doubt on whether the corporate ethics boom of the '90s has made any difference. A 2001 study found ethics programs tended to rely too heavily on asking individual employees to do the right thing and disregarded the impact of organizational culture on people's behavior. The authors of this study insisted that ethics programs would remain ineffective until companies developed the means to integrate ethical values into daily routines.[14]

The recent corporate scandals underscore that many of these programs were little more than window dressing. Enron broadcast its ethics code—Respect, Integrity, Community, Excellence (or RICE)—on everything from coffee mugs to T-shirts and workplace banners, while Kenneth Lay gave speeches at conferences on corporate ethics.[15] But when ethics really counted, Enron's board of directors waived its ethics policies to let Andrew Fastow engage in shady business deals. Elsewhere, the ethics compliance officer at Tyco is one of the corporate executives under indictment. The case of Arthur Andersen is especially ironic. Even as it made hundreds of millions of dollars in its ethics consulting division, Andersen refused to establish an internal ethics-compliance program—an omission that came back to haunt it later.[16]

Leaving aside all the scandals, the ethics at many companies remain dismal. A 2003 survey by the Ethics Resource Center found that 22 percent of employees said that they had witnessed misconduct either often or occasionally during the previous year. A large survey of corporate employees in 2000 by KPMG found that half of the respondents had observed violations of the law or company standards during the previous year, and many reported that these violations were quite serious. In yet another study, less than half of the employees surveyed felt that "their senior leaders are people of high integrity." Given all this bad news, who would believe that there's been a revolution in corporate ethics in the past decade?[17]

Still, the new ethics programs are far from worthless. These codes work better some places than others, and there is much to be learned from past successes. Nearly all experts agree that the defense industry offers the most instructive model for effective reform. Defense contractors were synonymous with dirty business dealings during the 1970s and 1980s. This began to change in 1986, with the initiation of Operation Ill Wind, a Defense Department sting operation against corrupt defense contractors that found pervasive illegal behavior. In response to the scandal, leading defense firms subscribed to the Defense Industry Initiative (DII), a sweeping program designed to "aggressively self-govern and monitor adherence to [the DII] code and to federal procurement laws."[18]

Businesses that are serious about creating an ethical climate would be wise to follow practices that have become common in the defense industry. These include regular employee training in ethics codes, more ways for employees to blow the whistle, and genuinely open ties to government regulators. Most of this is pretty basic stuff. Running an honest business is not rocket science, it turns out.[19]

The real challenge is that many business leaders just don't buy the idea that more integrity means higher profits. Apostles of socially responsible business promote this idea as gospel, but it has yet to be backed up with convincing research, according to Lynn Sharp Paine, a professor at the Harvard Business School and one of the world's leading experts on corporate ethics. But business leaders are slowly beginning to pay more attention to thinkers like Paine who advocate adding ethics to the bottom-line calculus. As Paine sees it, changing public attitudes around the world are forcing corporations in this direction. "Today's leading companies are expected not only to create wealth and produce superior goods and services but also to conduct themselves as 'moral actors'—as responsible agents that carry out their business within a moral framework…society has endowed the corporation with a moral personality." Paine backs up this assertion with survey data about the changed views of employees, citizens, consumers, and even some investors. This new outlook is diametrically opposed to the trend within corporations over the past two decades, which has been to strip companies of moral purpose and focus solely on profit maximization. Ordinary people, it appears, aren't so keen on that trend.

Paine is prodding corporate leaders to get in step with the public by pointing out ways that an investment in ethics pays. She emphasizes four points, all of which jibe with common sense: It helps companies to manage risk, especially the risk of a fatal scandal. It enhances the functioning of companies by building positive and cooperative values among employees. It strengthens the brand identity and market position of companies by enhancing or protecting their reputation. And it improves companies' "civic positioning," thus allowing them to better get along with communities and governments.[20]

Conservative commentators have often been right to point out the great inefficiencies incurred by government regulation. It *is* costly to have government constantly looking over the shoulder of business. Yet time and time again, even the most respected companies have shown that they cannot be trusted when it comes to the money of investors, the health or safety of consumers and workers, or the protection of the environment. A beefed-up commitment to ethics in the business world will not take away the need for vigilant government regulation in the near future. But it may not be too far-fetched to imagine that, over the next few decades, this commitment could become deeply institutionalized, that the notion of self-policing could actually come to have meaning, and that the need for intervention by government could diminish.

Such a shift toward stronger ethics is more likely to occur in a business world where today's conception of the bottom line has less influence. In so many professions, ethical problems have escalated in tandem with growing bottom-line pressures. Neither outside regulation nor internal ethics programs can force the private sector to abandon the extreme bottom-line thinking that it has embraced over the past two decades. However, both types of reform can curb the worst abuses associated with this shift and push companies to not measure performance by earnings alone. Despite the laissez-faire revolution of the '80s and '90s, the profit motive has never achieved total dominance in the business world and midcentury notions of responsibility to multiple stakeholders have not been extinguished entirely. Many of the reform challenges that face the private sector boil down to the task of reinventing these notions for a new era—and, here again, putting market values back in their proper place.

Teaching Integrity

There is nothing fantastical about the society I am sketching out—a place where people believe that the rules are fair and that if they play by the rules, they will get ahead; where people feel more connected to each other and less obsessed with piling up possessions; where business is serious about following both the letter and the spirit of the law.

We don't have to alter the DNA of Americans to create this kind of society. Most of us want it already. However, in addition to other sweeping reforms, we do need to alter the attitudes of young people. If the next generations of Americans are to help build a more ethical society and sustain it, they must come of age within institutions that are far less tolerant of cheating than today's high schools and universities. They also must learn early on to commit themselves to principles beyond their own individual self-interests.

One important step is character education in our schools that starts at a young age. Many children are not taught much about ethics and honesty at home. Parents are working longer hours than ever before and a lot of them don't have time for moral discussions. They may also lack the skills or vocabulary to have such talks in an effective way. Worse, many parents may be caught up in the cheating culture themselves and set a negative example for their children. Polling shows that parents are deeply worried about their ability to inculcate their children with strong values, while young people today are acutely aware of the moral crisis facing America.[21]

Educators in recent years have increasingly embraced the idea that character education should be part of school curriculums. Most character programs take into account the obvious potential for controversy in such work and steer clear of divisive issues. "Character programs are based on commonly held values—like respect, responsibility, fairness, honesty, justice—that we believe should be adopted by the community," says Esther Schaeffer, who runs the Character Education Partnership, a national nonprofit organization. Character programs tend to be flexible and are often shaped to fit local needs with the input of the community. By the late 1990s, public schools in almost all states promoted character education at some level, although this commitment is typically very small and made possible with the support of a pilot program sponsored by the federal Department of Education.[22]

These programs work in different ways, explains Schaeffer. The notion of establishing the legitimacy of authority is often very important. "In an elementary school class, it might be rules put up at the beginning of the year—but instead the teacher lets the kids help develop the rules of how to behave together. The kids usually come up with better rules, and they have developed them, so they feel more a part of it. Or perhaps the teacher will start the day with a classroom meeting, and students can talk about anything that troubled them the previous day, or they might share ideas about a specific topic. So there is a definite effort to have kids bring up and handle issues.... A school where you've got good character education is one where the culture of the school puts a high premium on respect, honesty, and kids being responsible for their actions and adults doing the same."

Character-education programs are still relatively new. But some research suggests that they have real impact, reducing cheating and other forms dishonesty among students. These programs work as a natural complement to honor codes, which are a more well-established strategy for reducing cheating.[23]

Honor codes in schools have been around for over a century. They range from simple edicts—"cadets will not lie, cheat or steal, or tolerate those who do"—to highly elaborate written codes supported by accumulated case history. Like character-education programs, many of these codes let students play a key role in enforcing the rules.

Research by Donald McCabe and other experts shows that honor codes do reduce cheating—with some caveats. Honor codes must go hand in hand with a strong institutional commitment to academic integrity. High schools and universities need to educate students and faculty about honor codes in an ongoing way, and use creative strategies to foster an environment where cheating is not socially acceptable. Some schools, such as Washington and Lee University, even discuss the campus honor code with students during the admissions process. Many schools include dialogue and ceremonies around the honor code as part of their orientation process for new students. At Vanderbilt, for example, each freshman signs a class banner affirming their commitment to the school's honor code. Banners for all of the classes enrolled at Vanderbilt hang in the school's student center for everyone to see. The larger point here is that students must be constantly reminded of their solemn obligation to academic integrity.[24]

A school's commitment to integrity must aim to change not just students' attitudes but also those of faculty. Right now, faculty are a big part of the cheating problem—with students commonly reporting that teachers and professors let cheating go on. Many teachers don't seem to think that policing integrity standards, and all of the hassles that entails, falls under their job description. But obviously it does. Schools must make it clear to faculty that keeping students honest is as important as their other obligations, such as teaching and publishing. Schools should measure whether faculty live up to that obligation, just as they measure performance in other areas.

Teaching young people to look beyond their own narrow self-interest is even more important over the long run than strengthening standards of academic integrity. Young people must be pushed to explore the world beyond their MP3 players and to develop themselves as more than consumers and future workers. Myriad programs exist that encourage young people to do exactly that—by helping the elderly or homeless, improving public spaces, and working in politics, the arts, or education. Many high schools and universities now have community service requirements. While there is much debate about exactly what approaches work best to elevate young people out of a self-centered mind-set, we have more knowledge about this than ever before, including many insights from the AmeriCorps program and such large-scale nonprofit organizations like City Year. Here, as elsewhere, it is not that we lack solutions to the fraying moral fabric of our society; it's that we lack the political will and resources to enact these solutions on a greater scale.

Finally, we must get serious about teaching ethics in business and professional schools. These programs represent the last chance educators get to shape the attitudes of future leaders. Right now this chance is largely squandered, particularly at business schools, where many students arrive with dishonest instincts and often leave in even worse shape. One 1998 study revealed practically no difference in the moral orientation of business students and convicted felons when making hypothetical decisions about ethics in the workplace. If anything, the study's authors contended, the results of their surveys showed more integrity among the felons, who were less likely than business students to raid competitors for desirable employees and to slight the needs of their customers in favor of stockholders.[25]

The Association of Advanced Collegiate Schools of Business requires accredited M.B.A. programs to educate students about ethics, allowing schools to choose their own method for integrating ethics into the business curriculum. The results of this requirement are not impressive. A study conducted by the Aspen Institute's Initiative for Social Innovation Through Business between 1999 and 2001 found that students left graduate business programs with a weaker ethical orientation than they had when they arrived.[26]

Getting serious attention for ethics has traditionally been an uphill battle in business schools. Business school faculties often resist ethics initiatives, and ethics requirements are frequently shunted to the sidelines—with students and professors alike simply going through the motions to get the requirement out of the way. For example, in the wake of the insider trading scandals in the late 1980s, former SEC chairman John Shad gave Harvard Business School a $20 million personal gift to establish an ethics program. Today most people agree that the school has largely wasted the money—and missed the opportunity to be a leader on teaching ethics.[27]

That's a shame. Creative ideas do exist that can make M.B.A.s really stop and think about ethics, but they need to be popularized and legitimized. Forays to federal prisons are one such idea. Class visits by corporate ethics officers and executives who've faced ethical dilemmas can also have an impact. Learning to stay honest in business, one would hope, is not all that hard. The real challenge is making business school a place that teaches such lessons.

Preprofessionals in law, accounting, and medicine also badly need stronger ethics training. A study of undergraduate accounting programs in 2000 showed that these programs failed abysmally to instill integrity. Accounting students were queried during their sophomore year and again as seniors about their willingness to falsify data on financial reports. The results showed that seniors were more willing than underclassmen to cheat on the balance sheets.[28] Students in all law schools are required to pass a course in professional responsibility, as well as to pass the Multistate Professional Responsibility Examination as part of most state bar exams. But these requirements provide future lawyers with little real-life guidance for dealing with ethical quandaries on the job, particularly around billing. Elsewhere, medical schools do little to prepare students for the ethical minefield that is modern medicine.[29]

In all of these fields, there exist plenty of ideas for how to teach ethics in ways that will have traction in real life. The problem is that teaching preprofessionals how to act honestly is nearly always an afterthought. It's past time for leaders within professional schools to put ethics front and center.

America's cheating culture will not be dismantled overnight. The reforms I've suggested are just part of what needs to be done, and any comprehensive agenda to curb cheating will take years to put into place—if our society makes a real commitment to take action.

That's a big "if." Some reforms, like improving honor codes in schools, are relatively easy because they will not encounter much opposition or because they can advance institutional self-interests. Others, like reducing the influence of money in politics, run up against powerful interests that will fight to the death to preserve the status quo.

Certainly, many of the reforms needed to curb cheating will be harder to enact as long as the United States remains so deeply in the grips of laissez-faire ideology and market values continue to reign with such influence in our culture and our economy.

How long will the present era of extreme capitalism last? That is hard to predict, but I'll offer a few encouraging thoughts. First, as the historian Zachary Karabell and others have argued, American history has been characterized by the rise and fall of particular ideas that dominate American life for a few decades before giving way to the next big idea. Religion dominated American life in the colonial era. Expansionism and the drive for national unity dominated U.S. society in the nineteenth century. Government played a dominant role in shaping American life during the mid-twentieth century. And, since the 1970s, the market has been the guiding force in American life. Each stage in American history has given way to something else and often the new stage is a backlash to what existed before.[30]

Given this record, the present hegemony of the market is unlikely to last indefinitely. There are already signs that a new big idea may be on the rise—a concept of "life balance," or what Karabell calls "connectedness." A drive toward balancing work and acquisition with stronger community and greater personal well-being is emerging amid growing public dissatisfaction with our consumerist, work-obsessed society. Signs of such a trend can be seen in the "downshifting" movement, the fight against sprawl, rising interest in eco-tourism, and much else. However, exactly when—or whether—these disparate trends will metastasize into a full-blown cultural shift remains to be seen.

A backlash to the market extremism now dominant in U.S. society could also take a more familiar, cyclical path, as has occurred several times in the past century. The early twentieth century saw a revolt against the excesses of the Gilded Age and the undemocratic power of the trusts. The 1930s saw a backlash to the laissez-faire economics of the 1920s and the lords of the financial sector. The 1960s saw a backlash to the consumerism and conformity of the 1950s, as well as the overreach of corporate power in U.S. society. Americans love the free market, but again and again they have risen in revolt when market forces became too powerful.

It is too early to say whether the United States is on the verge of one of these periodic revolts. But many of the conditions that have catalyzed such backlashes in the past are now present: growing public concerns about unchecked private power; high-profile scandals that underscore the corrupt role that corporations play in

America's politics; and mounting unease with the social and cultural consequences of extreme competition and materialism. It is worth adding that every past popular revolt against laissez-faire excesses has come on the heels of a Republican presidency in which the White House was widely seen as too cozy with big business. We can only hope that George W. Bush's presidency—a presidency made possible by money from corporations and the superwealthy, and largely geared toward serving these constituencies—will also go down in the history books as the last stand of a deeply corrupt social order.

Banishing the latest plague of market fundamentalism from U.S. society, forging a new social contract, reforming business and professions, teaching integrity to the young—these are the changes needed to dismantle the cheating culture. Some of us are positioned to be actively involved in such reform efforts; all of us can work as individuals, in our daily lives, to strengthen the moral fabric of U.S. society. Yes, the cheating epidemic writ large can be traced back to large-scale shifts in America over recent decades, but that doesn't mean that as individuals we're not responsible for our actions.

For what it's worth, my advice to those who cheat or are around cheating boils down to two simple ideas: one, go ahead and be a chump; and two, don't hesitate to be a pain in the ass.

Be the chump who files an honest tax return, even if you believe that "everybody else" is cheating. And be the person, also, who gives your friends a hard time for cheating on their taxes.

Be the chump who pays $18 for a CD at the store, even though record companies are monsters and you can get the music for free over the Internet. And when a friend sends you a file of a recent hit song, send it back to him with a lecture on copyright law.

Be the chump who tells Allstate that your boyfriend—the one with sixteen points on his license—does, in fact, live with you and does share your car. Auto insurance companies are a bunch of bloodsucking profiteers, but fraud isn't going to solve this problem and actually makes it worse.

Be the chump at work who doesn't tell that endless blizzard of white lies that helps people get ahead in business. And, whenever you get the chance, blow the whistle on corruption at the office.

If you're a lawyer, be the chump who only bills your actual hours. If you're a doctor, stay away from Big Pharma reps bearing gifts. If you're an athlete, stay clean even if it means a life of amateur sports.

All of this is easier said than done, I know. There are a lot of people who feel unable or are unwilling to stop cheating until broader structural changes are in place. And who am I, really, to lecture people about the sacrifices they should make to improve our society's ethical climate—a goal that's pretty abstract compared to providing for one's family.

In the end, one of the most important changes needed to reduce cheating in many places is to banish the perception that "everybody does it." If you don't cheat, the category of "everybody" will include one less person. And if you set an example, or persuade others not to cheat, than "everybody" could become "some people" and maybe, in time, "a few people." The tipping-point phenomenon can go both ways: cheating can become common enough to be normalized and socially acceptable; but it also can turn uncool, thanks to ordinary people who decide to take an active moral leadership role. Each of us has real power to help influence the climate of social norms that surrounds various kinds of cheating. Use that power.

A special word to parents. If you're a parent, don't wait for the educational system to adopt character-education programs or serious honor codes. Make a commitment to integrity in your own home. Talk to your kids about why they should play by the rules—and honestly challenge rules they think are wrong. Teach them how to work through the tough ethical dilemmas in life. Create an environment where money and status do not loom in your children's lives as the greatest good.

One last piece of advice to those who have read this far: beware of ATMs that appear to be dispensing free cash. In this electronic age, alas, nothing is ever forgotten.

afterword

Since publishing *The Cheating Culture* in January 2004, I have been reminded of just how personal this topic is. In my travels around the country, and through letters and e-mails, I have had many people pour their hearts out to me. I met an executive who moved his family across the country to take a dream job that paid a million dollars a year—only to have to quit shortly thereafter because his new bosses lacked integrity. I heard from a building contractor who wanted to run his business honestly but found that it was impossible in his industry because all of his competitors were breaking the rules. I met a college professor who had gone after a plagiarizing student—in an ethics class, no less—even though his institution made it clear that if he got sued by the student, he was on his own. Faculty are afraid to do the right thing, he said. I spoke with a leader in philanthropy who was hired to run a foundation set up by a recently deceased billionaire—and who watched helplessly as the foundation board of trustees instead used the funds to line their own pockets. I heard an investment banker, an entrepreneur, an accountant, and others who had been successful in business all say the same thing: that they could not have succeeded without cheating, and they were not proud of that fact.

The culture of cheating is not an abstract phenomenon. It is a very real part of people's lives. And most of us don't have a clue about how to deal with the tough ethical choices that come our way.

Since *The Cheating Culture* was published, Martha Stewart has been sentenced to prison. Ken Lay and Jeffrey Skilling have been indicted. A new steroid scandal has engulfed the sporting world. A top reporter at *USA Today*, Jack Kelley, resigned in another major case of journalistic fraud. Federal investigators launched a sweeping investigation of the illegal marketing practices involving pharmaceutical companies and doctors. The record industry has filed a flurry of lawsuits against music pirates. New revelations emerged about how large retail stores cheat employees out of deserved pay by manipulating computer time sheets. Eliot Spitzer launched probes of numerous mutual funds. And, in the political sphere, the Bush Administration found itself confronted by evidence that it lied about the threat posed by Iraq and lied about the cost of the new Medicare drug benefit—two whoppers even by Washington's standards.

In some areas, there are signs that we may be turning a corner and entering an era of reform. If twelve-year-old music pirates are being sued, it is safe to say that a new draconian spirit is in the air. But in other areas the clean-up crews have clearly just begun their work, and despite what prosecutors may declare, many deep-seated forms of corruption in American society have barely been disturbed.

Start with the business world. Over the past year, prosecutors have scored some high-profile white-collar convictions, and the tinkle of handcuffs against cuff links has been heard often outside of courthouses as perp walks by the Brooks Brothers set have continued. One hopes that the conviction of Martha Stewart on four felony counts, and her sentence in July 2004 to five months in prison, will send a useful message to the corporate elite—namely, that even the best defense lawyer and a few million dollars in PR will not save you from jail if you slip up at the office. On the other hand, the belabored Tyco trials of the past year would seem to send the opposite message—that corporate fraud cases add up to a swamp of mind-numbing ambiguities, even when someone apparently steals a few hundred million dollars from his own company. Dennis Kozlowski emerged as the poster pig for corporate greed at the height of the corporate scandals. But when it came time for a jury to sort through his alleged crimes, it turned out that things were pretty darn complicated and that, in fact, a halfway coherent argument could be made that much of Kozlowski's bad behavior was technically legal. The six-month trial of Kozlowski and Tyco's former CFO Mark Swartz ended in a mistrial in April 2004. Maybe prosecutors will nail these two in the retrial that starts in early 2005. Maybe not. In July 2004, Mark Belnick, Tyco's former general counsel, was acquitted by a jury on fraud and larceny charges. A vast fortune was siphoned out of Tyco by top executives during the boom, but so far prosecutors have yet to prove that any of them committed a single crime.

Developments in other corporate cases are more reassuring. Prosecutors won convictions against two members of the Rigas family in the Adelphia case in July 2004, and these men will probably do some real jail time. The second trial of investment banker Frank Quattrone also ended in conviction in May 2004, and he seems bound for at least a few months in a federal prison. Meanwhile, after a long delay, the government finally is making some progress in bringing to justice the high command at both Enron and WorldCom. Kenneth Lay was indicted on eleven felony charges in July 2004. Jeffrey Skilling, indicted earlier in the year, faces three times as many charges. Both men maintain their innocence and blame the fall of Enron on Andrew Fastow, the former chief financial officer who pleaded guilty to two felonies and accepted a ten-year prison term in January 2004. Fastow will be testifying against his former bosses, and presumably he has an altogether different take on things. The trials of Lay and Skilling are sure to be prolonged and complicated. Judging from the Tyco trials, convictions are anything but guaranteed.

The WorldCom case is also moving along. In March 2004, Scott Sullivan pleaded guilty to federal charges and agreed to cooperate with prosecutors against Bernard Ebbers, the fallen founder of WorldCom. In entering his plea, Sullivan explained himself to a federal judge this way: "I took these actions, knowing they were wrong, in a misguided effort to preserve the company to allow it to withstand what I believed were temporary financial difficulties." That explanation sounds plausible, although it's hard to think that Sullivan wasn't also motivated by the vast fortune he stood to earn as long as the company's stock remained high. Funny how people don't think that their own greed passes as much of an excuse for multiple felonies—when, in fact, such a motivation makes perfect sense in this day and age. Will Ebbers soon be running his Sunday bible classes in a federal prison because of Sullivan's testimony? That depends on whether such testimony, along with other evidence, will amount to a smoking gun. The trial will probably last well into 2005.

As the government has moved against the biggest corporate villains over the past year, many in the public seem to think that justice is being done. It isn't. The ultimate tally of convictions and jail terms is unlikely to be impressive enough to deter new financial crimes a few years from now, when the hype has died down. Consider that corrupt corporate insiders pocketed tens of billions of dollars during the boom and that hundreds of companies, executives, stock analysts, and accountants were investigated for wrongdoing. Pension funds lost hundreds of billions of dollars in investments made on the basis of falsified financial data. Yet, five years from now, you'll be able to count on one hand the number of corporate bad guys from this period who are still in prison. With the benefit of hindsight, and the help of a calculator, future would-be wrongdoers may judge that crime paid for most of those involved in boom-time shenanigans during the late 1990s. And while some of the rules around accounting have been tightened, and the penalties for corporate fraud have been strengthened, plenty of loopholes still exist that allow executives to mislead and cheat investors without doing anything illegal.

Major new corporate scandals within the next decade are inevitable.

In October 2003, the U.S. Anti-doping Agency announced that it had identified a new designer steroid known as THG. The steroid had been invisible to standard drug tests until it was unmasked by a UCLA scientist named Donald Catlin. Once the properties of THG were known, the steroid was traced to the Bay Area Laboratory Co-operative (BALCO) in California and then to a number of top athletes. The result was the largest drug scandal ever to hit the sports world. A federal grand jury eventually indicted several people associated with BALCO on money laundering and other charges, and numerous athletes worldwide have been disciplined or investigated for using THG. Among those alleged to have received the drug is Barry Bonds, whose personal trainer, Greg Anderson, was among those indicted in the BALCO case. Marion Jones, the 2000 Olympian, has also been mired in the THG scandal—although no definitive judgment about whether she used the steroid has yet been issued.

For months the scandal dominated the sports pages and occasionally the front pages. President Bush condemned steroids in sports in his January 2004 State of the Union Address. The Senate held hearings on the issue, in which Senator John McCain declared that baseball was in danger of "becoming a fraud in the eyes of the American people."

The THG flap has had two effects. First, it has reminded athletes that the science of antidrug testing never rests. A drug that is undetectable today may be easy to pick up tomorrow. And second, the scandal has brought tremendous pressure to bear on sports authorities—especially Major League Baseball—to get more serious about drug-testing. The players union has continued to block serious action, and the MLB's drug-testing policy, implemented in 2003, is widely viewed as a joke. This laxity cannot endure. McCain and other senators have made it clear that if the MLB doesn't crack down on steroids, Washington will step in to save the hallowed American institution of baseball. Big government is good for some things, it turns out.

New victories in the battle against drugs in sports are likely to be fleeting, though. While the ethical climate in sports should improve in the short run as a result of the THG scandal, the incentives for athletes to cheat will remain powerful. Winner-take-all trends in athlete pay are sure to keep accelerating as the reach of global media expands, creating bigger audiences for televised sporting events as well as product advertisements that feature star athletes. Moreover, the scientific arms race between dopers and drug-testers ultimately favors dopers. Experts predict that it is only a matter of time before athletes start to make use of genetic enhancements to improve their performance. Once these technologies are perfected, the battle for clean athletic competition may be lost for good.

In May 2004, Warner-Lambert agreed to pay $430 million to resolve criminal and civil charges related to illegal marketing of the drug Neurontin by its subsidiary Parke-Davis. More than half that money went to pay a criminal fine to the federal government and more than $100 million went to state governments. In a statement, the Department of Justice said that the illegal marketing of Neurontin had endangered the health of patients, compromised the integrity of doctors, and led to $150 million in losses to federal and state health-care programs. As part of the settlement, Warner-Lambert pleaded guilty to two felonies.

Predictably, not a single individual involved in these crimes was held accountable, charged with anything, or even mentioned by name by the Justice Department. The financial pain of the settlement will be borne not by the executives who sanctioned the crimes or the managers who implemented it or the doctors who benefited from it financially—but by the shareholders of Pfizer, Warner-Lambert's parent company, who did nothing wrong. The settlement is another example of the hypocrisy in America today around the concept of personal responsibility. Steal $1,500 from a bank in an unarmed robbery and you're apt to spend years in prison. Steal $150 million from Medicaid and Medicare and your name doesn't even appear in a court document. It is hard to see how this kind of settlement will have much deterrent value at all on the pharmaceutical industry—or on the doctors that pocket tainted money.

Still, it is reassuring to see that the Feds are finally targeting corrupt pharmaceutical firms in a big way. Nearly every large drugmaker in the United States has now received subpoenas by federal investigators probing illegal marketing practices and other possible crimes. In July 2004, Schering-Plough agreed to pay a $350 million settlement after a six-year probe by federal investigators found that it overcharged Medicaid for drugs. Again, no individuals in the company were charged with anything, even though Schering-Plough has been in trouble with the law before; in 2002 it agreed to pay $500 million to settle charges by the FDA that it was dangerously lax about quality control at its drugmaking facilities. Crimes keep occurring at Schering-Plough, but somehow nobody is ever blamed.

IRONICALLY, music pirates have more to fear these days than pharmaceutical executives who have orchestrated large-scale frauds: Over the past year, the Recording Industry Association of America has filed lawsuits against several thousand of them. By the middle of 2004, the RIAA had settled about seven hundred cases, with some defendants paying as much as $15,000, not to mention legal fees—personal penalties greater than any imposed on the scam artists at Warner-Lambert or Schering-Plough. It is not clear whether the RIAA's campaign of fear is working. An opinion survey taken in late 2003 by the Pew Internet Project found that music piracy had fallen sharply as a result of the RIAA's first round of lawsuits. However, in July 2004, BigChampagne, a technology market research and consulting firm, issued more authoritative data that found that file swapping has increased nearly 20 percent in 2003–2004. Meanwhile, the Motion Picture Association of America warns that the online piracy of movies is growing rapidly.

Both the music and movie industries are facing an uphill fight as broadband Internet connections spread, making it easier to download large files. Ever more draconian legal action is always an option. The spectacle of Internet pirates going to prison—an extreme solution, granted—would surely have an impact. Legal alternatives to piracy, such as iTunes, also offer hope. Paid downloads from iTunes have swelled over the past year. The moral of the iTunes story is that there are many people out there who prefer to do the right thing, if given an option. More efforts by the entertainment industry to co-opt the online search for music and films may be best solution to electronic piracy.

THE EPIDEMIC OF student cheating has been in the news frequently since *The Cheating Culture* was first published. While no single event has catalyzed this new attention, academic dishonesty has been the subject of front-page stories in major papers, and ABC aired a two-hour television special about the problem. After years of simmering below the surface, a quiet plague at both high schools and colleges, rampant student cheating is finally drawing attention from the likes of Charles Gibson. That by itself is a step forward. What remains to be seen is whether all this new concern will translate into more honor codes, tougher penalties for cheating, more support for faculty who try to punish cheaters—and ultimately less cheating. We should have some answers soon, as survey data by researchers like Donald McCabe and Michael Josephson become available about cheating trends among students in the first years of the twenty-first century, and as new research is published by the scholars who track institutional policies toward cheating.

Dismantling the culture of cheating that exists in many academic settings will be pick-and-shovel work, given how entrenched this culture is. It can take several years for a university to put into effect an honor code and new set of procedures around academic integrity, and years beyond that for these changes to reshape student attitudes. However, there is always the possibility that progress will occur much faster than anyone expects. Norms among teenagers are famously unstable. New fads and values can stampede into youth culture practically overnight. What is cool now can be very uncool next week, and the dominant youth values of today can be anachronisms a few years down the line. While I have argued that the cheating culture in schools reflects deep-seated social and economic conditions, the scope of the problem also reflects the self-reinforcing dynamic of cheating. Kids cheat because other kids cheat, and it is seen as acceptable. If this dynamic can be disrupted—if a new climate of peer disapproval can be fostered around cheating—than the pendulum could swing dramatically in the opposite direction.

Over the past year, I've met or heard from quite a few people who are taking on the cheating culture in new and creative ways. These ethics entrepreneurs aren't waiting for a national organization to tell them what to do, or for some senator to introduce legislation that they can work to pass. They are just moving ahead, often quite alone, to do whatever they can to make a difference. Consider a few examples:

Nomi Prins, an investment banker steeped in the arcane details of structured debt and derivatives trading, decides she has had it working for a corrupt industry. She quits her lucrative job to become a crusader for banking reform. Now she advises policy makers, pension funds, and think tanks about how to reform a financial world that remains largely unchanged since the corporate scandals. She writes for business magazines, exposing financial rip-offs, and is the author of a book, *Other People's Money*, that sheds new light on how the corporate scandals happened—and why new ones are inevitable.

Frank Shorter, an Olympic gold medalist from the 1970s, a star in his day, grows concerned about the epidemic of doping during the 1990s. He is appalled by how many gaps there are in enforcement efforts and becomes active in anti-doping efforts worldwide. He testifies before Congress and speaks widely on the topic. Working with others, he helps to found a new organization—the U.S. Anti-doping Agency—that becomes the leader in the fight to reduce doping among U.S. athletes.

Rick Bishop, a veteran builder, sees that the construction industry is badly in need of ethics reform. He decides to do something about it. He starts a new business, Integribuild, to provide ethics training to building companies and to bring their ethics standards into the twenty-first century. He offers to give free trainings through a leading trade association, but they aren't interested. In fact, he finds that nobody in the industry is much interested in ethics. He isn't discouraged. He thinks it is only a matter of time before the construction industry gets on the ethics train. And he's right.

Walt Pavlo, a former business executive convicted of financial fraud, steps out of prison into his new life. During the 1990s, he had gotten caught up in the culture of greed. He had gone along with a colleague who had a bright idea to make them both rich, and he had grabbed for money that wasn't his. He got busted and did time. Now he works full-time trying to explain to others just how slippery the slippery slope can be. He talks to managers around the country and has a simple message: If it can happen to me, it can happen to you. But he also knows how critical it is to reach young people, and so he has devoted much of his effort to working with teenagers and young adults just starting out in business.

I have met many other ethics entrepreneurs beyond these, and I find their efforts not just impressive, but reassuring. The cheating culture is being contested every day, with growing zeal and creativity. Beyond those challenges mounted by a proliferating array of ethics entrepreneurs, a new progressive reform movement has palpably gained steam since I finished *The Cheating Culture*. This movement is not unlike the reformist upsurge of a century ago. It is led by groups like MoveOn.org and the Center for American

Progress. It is taking on the growing dominance of corporate power over U.S. society by pushing for new rules around media ownership and campaign donations, and new protections of consumers, patients, and investors. It is addressing the economic insecurity of Americans by working for affordable health care, a living wage, and other policies that can ensure a decent life for all. It is trying to fulfill the promise of our democratic heritage, and get more people to vote and participate in civic life. The new progressive reform movement is also challenging the right wing's monopoly on moral language and broadening the discussion of values in America, arguing that the harsh values of the market have gone too far and it is time to reassert the ideals of community and fairness.

The new energy of the progressive movement gives all of us more choices for taking on the cheating culture. You can join an organization or support a candidate fighting for a fairer, more democratic America. You can be an ethics entrepreneur, challenging corruption and complacency in your own little corner of the world. Or you can work for better values more privately—in the way you raise your children and how you handle situations at work or in your financial life. Nearly every day, you can choose whether you are part of the cheating culture or against it.

Integrity, as the famous saying goes, is what you do when no one is looking.

sources

Cheating is a difficult subject to research. With the exception of the large body of research on cheating by students, there is a lack of data about most forms of cheating—but especially data that allow for comparisons across time. For example, given the lack of comprehensive or effective drug testing in professional sports, it is impossible to say for sure how many athletes currently use performance-enhancing drugs—much less reliably compare rates of such use today to past periods. Likewise, while some forms of white-collar crime are carefully tracked, many other incidents of unethical or illegal conduct in business go undiscovered or are dealt with internally by companies and not recorded in any public place. Reliable data on ethical misconduct within specific professions is also difficult to get for various reasons. State bar associations are in charge of disciplining lawyers, but most of them do not keep good records that date back in time and are easily accessible to the public. Much the same can be said of state medical societies and licensing boards, which are in charge of disciplining doctors. In such cases where data can be found that document the incidence of misconduct over time, that data must be evaluated carefully: after all, it can be hard to say whether people are cheating more often or whether their behavior is just being discovered more frequently due to better enforcement efforts, more transparency, or other reasons.

Survey research in which respondents self-report on their ethics is very scarce, again with the exception of education. There are few surveys of lawyers, doctors, professional athletes, or accountants that have asked the same questions about ethics over time. Surveys on the ethics of corporate employees by several organizations, including the Ethics Resource Center, do allow for comparisons over time, but most of these surveys only began a few years ago. Throughout the book, I have mainly used opinion-survey data in an effort to understand the public's views about personal values, job satisfaction, economic security, social trust, perceptions of fairness, and the like. Here and there, I have found surveys on Americans' attitudes toward different forms of cheating, such as auto-insurance fraud, music piracy, or tax evasion. Typically, though, these polls are one-shot deals, and don't allow for serious comparisons over time. For example, over the past thirty years, the General Social Survey has only twice asked a question about cheating on taxes, in 1991 and 1998. In all cases where I used polling data, I approached such data with caution. Many polls have small sample sizes and outcomes can be strongly affected by how questions are posed. In most cases, my conclusions about public-opinion trends are drawn from multiple polls across a number of years.

Given these limitations on available quantitative data, I necessarily have relied heavily on qualitative evidence to support my claims. My lead research assistant for this book, LeeAnna Keith, did an extraordinary job of unearthing a voluminous quantity of secondary and primary source material on the various issues explored in this book, much of it dating back decades. Thus, for example, my exploration of integrity in the accounting profession not only draws on many contemporary sources but also on government studies of accounting undertaken in the wake of the corporate scandals of the 1960s and early 1970s, as well as the numerous articles written on the accounting profession in the early 1990s, when the large accounting firms were under fire for their role in the savings and loan scandals. Similarly, my analysis of ethics in the legal profession draws on a host of books, articles, and studies dating back to the 1970s that describe the major changes to the profession over the past thirty years. Some of the best sources for the book were found in specialized magazines, newsletters, and journals, or on industry-specific Web sites. Readers who want to dig deeper into the subjects covered in the book will find that the notes provide many leads for further exploration. I also suggest visiting my Web site at www.cheatingculture.com, which contains much additional source material on cheating. One comment about the citations: in many cases, full citations—especially accurate page numbers—were not available for sources accessed through electronic databases or over the Internet.

Extensive interviewing proved critical to this book. My research assistant Carolyn Rendell conducted and transcribed most of these interviews, bringing exceptional persistence, skill, and accuracy to this large undertaking. Most of the names of those interviewed are cited directly in the text, although a number of people requested anonymity. Those interviewed or consulted for this project include the following: Frankie Andreu, manager, U.S. Postal Service Cycling Team, 2000–2002, USPS racer, 1998–2000; Gerard Bisceglia, CEO, USA Cycling; Dr. Carlo Brugnara, professor of pathology, Children's Hospital and Harvard Medical School; Deborah Briggs, executive director, New York Huntington Learning Center; Jane Brown, vice president of enrollment and college relations, Mount Holyoke College; James E. Delany, commissioner, Big Ten Conference; William C. Dowling, professor of English, Rutgers University; James Duderstadt, president emeritus and university professor of science and engineering, University of Michigan; William F. Ezzell, chairman (2002), American Institute of Certified Public Accountants, and partner, Deloitte & Touche; Stuart C. Gilman, president, Ethics Resource Center; Robert Hirshon, president (2002), American Bar Association, and partner, Drummond, Woodsum & MacMahon; Dr. Sheldon Horowitz, director of professional services, National Center for Learning Disabilities; Arnold Huberman, Arnold Huberman Associates; Cecil Ingram, athletic director, University of Alabama, 1980–90; Marty Jemison, U.S. Postal Service racer, 1995–2000; John Kendzior, manager, employment services, Harvard University; Lisa Lerman, director, law and public policy program, Columbus School of Law, Catholic University of America; Stephen Loeb, professor of ethics, Smith School of Business, University of Maryland; Johnny Majors, head football coach, University of Pittsburgh, 1993–2001; Dr. Jim Martinoff, professor of finance, Graziadio School of Business and Management, Pepperdine University; Dominic Morandi, Schneider Legal Search; Randall Neal, managing director, The Broadmoor Group; Dr. Stephen J. Nicholas, director, Nicholas Institute of Sports Medicine and Athletic Trauma; Tom Parker, dean of admissions and financial aid, Amherst College; Gary Pavela, director, judicial programs, University of Maryland; David Pentlow, Katten, Muchin, Zavis & Rosenmen; Edward S. Petry, executive director, Ethics Officers Association; Dr. Jeff Podraza, PharmD manager, Drug Reference Line, U.S. Anti-doping Agency; Monica Ronan, senior human resources specialist, *TV Guide*; Chris Rugutsky, guidance counselor, Rudolph Steiner School; Dr. Allen Sack, professor of sociology and director of the Management of Sports Industries Program, University of New Haven; Esther F. Schaeffer, executive director and CEO, Character Education Partnership; Frederick A. O. Schwarz, Jr., senior counsel, Brennan Center for Justice, and chair, City Campaign Finance Board; Dr. Scott Sherman, assistant professor of management, Graziadio School of Business and Management, Pepperdine University; Thomas Siciliano, general counsel, Municipal Credit Union of New York; Stephen M. Voltz, managing director, QuantiLex LLP; Joseph T. Wells, founder and chairman, Association of Certified Fraud Examiners; David B. Wilkins, director, Program on the Legal Profession, Harvard Law School; Charles Yesalis, professor of epidemiology, Penn State; Gail Kenney, director of financial crimes, Internal Revenue Service.

Finally, during the process of writing and revising, I received generous help from various people who read all or parts of the book. Thanks to Tom Bonnett, Lew Daly, Tamara Draut, David Cay Johnston, Jessie Klein, James Lardner, David Smith, and Eric Uslaner. Thanks also to Leslie McCall for sharing with me her considerable knowledge about the different dimensions of economic inequality in the United States.

endnotes

Chapter One: "Everybody Does It"

1. John Cassidy, "The Investigation," *New Yorker,* 7 April 2003, 54.
2. Ibid., 58.
3. Robert Merton, *Social Theory and Social Structure* (New York: Free Press, 1957), 146.
4. Charles R. Geist, *Wall Street: A History* (New York: Oxford University Press, 1997), p. 64.
5. Zachary Karabell, *A Visionary Nation: Four Centuries of American Dreams and What Lies Ahead* (New York: HarperCollins, 2001), 122.
6. David R. Simon, *Elite Deviance,* 6th ed. (Boston: Allyn & Bacon, 1999). Simon's work, in turn, draws heavily from C. Wright Mills, who described the phenomenon of "higher immorality" among American power elites. C. Wright Mills, *The Power Elite* (New York: Oxford University Press, 1999).

Chapter Two: Cheating in a Bottom-line Economy

1. Michael Stumpf and auto mechanic quoted in Julia Flynn, "Did Sears Take Other Customers for a Ride?" *Business Week,* 3 August 1992, 24. For more on the problems in Sears auto repair shops, see Michael Santoro and Lynn Sharp Paine, "Sears Auto Centers (A)," Case Study 9-394-009, Harvard Business School; David Streitfeld, "Avoiding the Shaft," *Washington Post,* 22 June 1992, B5; Jane Applegate, "Paying Commission Can Backfire,"

Chicago Sun-Times, 23 September 1992, 60. On the tire balance allegations, see Joseph B. Cahill, "Sears Faces Another Suit on Auto Work," *Wall Street Journal,* 17 June 1999, A3; and Caroline E. Mayer, "Sears Fraud Case Keeps Rolling Along," *Washington Post,* 18 July 1999, H1. On New Jersey allegations, see Miles Moore, "N.J. Charges Sears with Tire Fraud," *Tire Business,* 28 October 2002, 3.

2. "The AM 100," *American Lawyer* (July 2003).
3. Karen Dillon, "Dumb and Dumber," *American Lawyer* (October 1995): 5–6; and Center for Professional Responsibility, *Annotated Model Rules of Professional Conduct* (Chicago: American Bar Association, 1999), 46.
4. Michael Trotter, *Profit and the Practice of Law: What's Happened in the Legal Profession* (Athens: University of Georgia Press, 1997), xv. Other studies of changes in the legal profession echo Trotter's conclusions. See, for example, Mary Ann Glendon, *A Nation under Lawyers: How the Crisis in the Legal Profession Is Transforming American Society* (New York: Farrar, Strauss, & Giroux, 1994); Anthony Kronman, *The Lost Lawyer: Failing Ideals of the Legal Profession* (Cambridge, Mass.: Harvard University Press, 1993); Marc Galanter and Thomas Palay, *Tournament of Lawyers: The Transformation of the Big Law Firm* (Chicago: University of Chicago Press, 1991); and Lisa Lerman, "The Slippery Slope from Ambition to Greed to Dishonesty: Lawyers, Money, and Professional Integrity," *Hofstra Law Review* 30, no. 3 (spring 2002): 883.
5. Richard Lacayo, "Tremors in the Realm of Giants," *Time,* 7 December 1987, 58–59.
6. "The AM 100," *American Lawyer* (July/August 1995): 55; and "The AM 100," *American Lawyer* (July 2001): 173.
7. Macklin Fleming, *Lawyers, Money, and Success* (Westport, Conn.: Quorum Books, 1997), 36. For Lerman's findings, as well as more on the problem of overbilling, see Lerman, "The Slippery Slope from Ambition to Greed "; and Lerman, "Blue-Chip Bilking: Regulation of Billing and Expense Fraud by Lawyers," *Georgetown Journal of Legal Ethics* 12, no. 2 (winter 1999): 205–365; Lerman et al., "Unethical Billing Practices," *Rutgers Law Review* 50, no. 4 (summer 1998): 2153–226; and William G. Ross, "The Ethics of Hourly Billing by Attorneys," *Rutgers Law Review* 44 (fall 1991): 13; William G. Ross, *The Honest Hour: The Ethics of Time-Based Billing by Attorneys* (Durham, N.C.: Carolina Press, 1996); and Andre Gharakhanian and

Yvonne Krywyj, "The Gunderson Effect and the Billable Mania: Trends in Overbilling and the Effect of New Wages," *Georgetown Journal of Legal Ethics* 14 (summer 2001): 1001–18.

8. Patrick J. Schiltz, "On Being a Happy, Healthy, and Ethical Member of an Unhappy, Unhealthy, and Unethical Profession," *Vanderbilt Law Review* 52, no. 871 (1999): 807.
9. On Helmsley overbilling, see Darlene Ricker, "Greed, Ignorance, and Overbilling," *ABA Journal* (August 1994): 63. On Spiotto's billing, see Lerman, "The Slippery Slope from Ambition to Greed," 905. On L.A. health-care company, see Amy Stevens, "As Some Clients Grow Bill Savvy, Others May Find They Get the Tab," *Wall Street Journal*, 11 February 1994, B5. On FDIC overbilling, see Linda Himelstein, "FDIC Counsel Routinely Overbill, Audit Finds," *New Jersey Law Journal*, 10 February 1992, 4. On Webster Hubbell's problems and rationales, see Ellen Joan Pollock, "Hubbell Receives 21-Month Prison Sentence for Bilking His Law Firm and Clients," *Wall Street Journal*, 29 June 1995, B4; and Adam Liptak, "Stop the Clock? Critics Call the Billable Hour a Legal Fiction," *New York Times*, 29 October 2002, G7.
10. Schiltz, "On Being a Happy, Healthy, and Ethical Member of an Unhappy, Unhealthy, and Unethical Profession," 807.
11. For a good summary of the Chicago School's rise, see Daniel Yergin and Joseph Stanislaw, *The Commanding Heights: The Battle for the World Economy* (New York: Simon & Schuster, 1998/2002), 123–31.
12. Charles Murray, *Losing Ground: American Social Policy, 1950–1980* (New York: Basic Books, 1995). See also Myron Magnet, *The Dream and the Nightmare: The Sixties Legacy to the Underclass* (New York: William Morrow, 1993).
13. Paul Krugman, "For Richer," *New York Times Magazine*, 20 October 2002, 62. Abrams quoted in Robert Reich, *The Future of Success* (New York: Vintage Books, 2002), 71.
14. "Executive Pay: Up, Up and Away," *Businessweek Online*, 19 April 1999.
15. Robert Kuttner, *Everything for Sale* (New York: Knopf, 1996), 3. For an excellent overview of the triumph of market ideas in American society during the 1980s and 1990s, see Zachary Karabell, *A Visionary Nation: Four Centuries of American Dreams and What Lies Ahead* (New York: HarperCollins, 2001), 119–57. See also Thomas Frank, *One Market Under God:*

Extreme Capitalism, Market Populism, and the End of Economic Democracy (New York: Doubleday, 2000); and David Bollier, *Silent Theft: The Private Plunder of Our Common Wealth* (New York: Routledge, 2002).

16. "Americans Express Little Trust in CEOs of Large Corporations or Stockbrokers," The Gallup Organization, 17 July 2002.
17. Cummiskey's story is related in Patrick Kiger, "What's Your Doctor Selling?" *Good Housekeeping,* 1 January 2001, 53. Cummiskey's husband also wrote a detailed complaint to AMA president Thomas Reardon in October 1999.
18. Benedict Carey, "A Supplemental Pitch," *Los Angeles Times,* 26 August 2002, A1.
19. Bradford H. Gray, *The Profit Motive and Patient Care* (Cambridge, Mass.: Harvard University Press, 1991). See also Marc A. Rodwin, *Medicine, Money, and Morals: Physicians Conflict of Interest* (New York: Oxford University Press, 1993). Wallace Simpson quoted in Carey, "A Supplemental Pitch," A1.
20. "Sale of Health-Related Products from Physicians' Offices," Code of Ethics, American Medical Association.
21. Diane M. Gianelli, "Ethics Council Revisits Office-based Product Sales," *Amednews.com,* 7 June 1999.
22. Ford Fessenden and Christopher Drew, "Bottom Line in Mind, Doctors Sell Ephedra," *New York Times,* 31 March 2003, A8.
23. On Franklin's experiences, see Liz Kowalczyk, "Parke-Davis Hired Firms to Strategize Drug Promotions for Unapproved Uses," Knight Ridder/*Tribune Business News,* 8 November 2002, 1; Liz Kowalczyk, "Pharmaceutical Whistle-Blower Describes High-Pressure Sales Tactics," Knight Ridder/Tribune News Service, Washington, D.C., 12 March 2003; and Melody Petersen, "Whistle-Blower Says Marketers Broke the Rules to Push a Drug," *New York Times,* 14 March 2002, C1. John Ford quoted in "Big Brother's Little Helper," *Harper's* (March 2003): 26.
24. For the *Times*'s major investigative study of this problem, see Kurt Eichenwald and Gina Kolata, "Drug Trials Hide Conflicts for Doctors," *New York Times,* 16 May 1999, 1. For more on this subject, including a look at the role of university medical researchers in clinical trials, see Pilar N. Ossorio, "Pills, Bills and Shills: Physician-Researcher's Conflicts of Interest," *Widener Law Symposium Journal,* no. 1 (2001): 75–103.
25. "Consumer Groups Accuse Makers of Epilepsy Drug Neurontin of End

Run Around FDA Regulations," *PR Newswire*, 4 February 2003. The details of Parke-Davis's behavior are laid out in the complaint *Congress of California Seniors, California Public Interest Research Group, U.S. Action v. Pfizer, Inc., and Parke-Davis.*

26. On speaker fees for doctors, see Melody Petersen, "Court Papers Suggest Scale of Drug's Use," *New York Times*, 30 May 2003, C1. On article honoraria, see Jennifer Heldt Powell, "Memo Details Parke-Davis's Push for Sales," *Boston Herald*, 30 October 2002, 31. Franklin quoted in Petersen, "Whistle-Blower Says Marketers Broke the Rules to Push a Drug."
27. "Consumer Groups Accuse Makers of Epilepsy Drug Neurontin of End Run Around FDA Regulations."
28. I say "seems" because there is little hard evidence of either an increase or decrease in conflicts of interest among doctors. According to some observers, today's conflicts are not necessarily more common than in the past, just different. There is a considerable literature on conflicts of interest in the medical profession and attempts to regulate these conflicts. See, for example Marc A. Rodwin, "The Organized Medical Profession's Response to Financial Conflicts of Interest: 1990–1992, *Milbank Quarterly* 70, no. 4 (1992): 703–41. See also Marc A. Rodwin, "Conflicts of Interest and Accountability in Managed Care: The Aging of Medical Ethics," *Journal of the American Geriatric Society* 46 (1998): 338–41; and Jerome Kassirer, "Financial Conflicts of Interest: An Unresolved Ethical Frontier, *American Journal of Law and Medicine* 27, nos. 2 & 3 (2000): 149–62.
29. David C. Colby, "Doctors and Their Discontents," *Health Affairs* 16, no. 6 (November/December 1997): 112.
30. Michael E. Debakey and Lois DeBakey, "Should Physicians Unionize? Yes, It Would Curb HMO Abuse," *Wall Street Journal*, 7 July 1999, A22.
31. Peter A. Setness, "Are Medical School Graduates in the Red over Their Heads?" *Postgraduate Medicine*, 20 December 2000, 13; and Robert Lowes, "More Hours, More Patients, No Raise?" *Medical Economics*, 22 November 2002, 76.
32. Setness, "Are Medical School Graduates in the Red over Their Heads?"; and "National Survey of Physicians," The Kaiser Family Foundation, May 2002.
33. "Healthcare Executives Most Highly Compensated," *Managed Care*, May 1999, 17; and "Managed Care's Profits Come from Physician Pockets," *Amednews.com*, 2 September 2002.

34. Ford Fessenden and Christopher Drew, "Bottom Line in Mind, Doctors Sell Ephedra," *New York Times*, March 31, 2003, A8; and Abigail Zuger, "Doctors' Offices Turn into Salesrooms," *York Times*, 30 March 1999, F1.
35. Neil Swidey, "AstraZenica's Huge Marketing Effort Aims to Keep 'Purple Pill' Profits Flowing," Knight Ridder/*Tribune Business News*, 19 November 2002; and Lewis Krauskopf, "Ask Your Doctor," *Bergen County* (N.J.) *Record*, 8 September 2002, B1.
36. David Hilzenrath, "Healing vs. Honesty?" *Washington Post*, 15 March, 1998, H1.
37. Leah Garnett, "Doctors Say They Deceive Insurers to Help Their Patients," *Boston Globe*, 12 April 2000, A1; and Donald A. Young, "Lying Unethical, Unnecessary," *USA Today*, 11 November 1998, A24.
38. Barry Bluestone and Bennett Harrison, *Growing Prosperity: The Battle for Growth with Equity in the 21st Century* (Boston: Houghton Mifflin, 2000), 30–31.

Chapter Three: Whatever It Takes

1. Quoted in Eric Uslaner, *The Moral Foundations of Trust* (New York: Cambridge University Press, 2002), 9.
2. Lawrence Mishel, Jared Bernstein, and Heather Boushey, *State of Working America, 2002–2003* (Washington, D.C.: Economic Policy Institute, 2002), 196 and 161. See also Holly Sklar, Laryssa Mykyta, and Susan Wefald, *Raise the Floor: Wages and Policies That Work for All of Us* (Boston: South End Press, 2003).
3. Graef Crystal, "Pay Hall of Shame and Fame—Class of 2001," *Bloomberg News*, 12 September 2001.
4. Data on after-tax income is from *Historical Effective Tax Rates, 1979–1997: Preliminary Edition*, Congressional Budget Office, May 2001. See also Isaac Shapiro, Robert Greenstein, and Wendell Primus, *Pathbreaking CBO Study Shows Dramatic Increases in Income Disparities in 1980s and 1990s: An Analysis of the CBO Data*, Center for Budget and Policy Priorities, May 2001.
5. On decreasing net worth, see Edward N. Wolff, "Recent Trends in Wealth Ownership, 1983–1998," Jerome Levy Economics Institute, April 2000. As of 1998, the top 20 percent of households owned more than 85 percent of all stocks. Mishel, Bernstein, and Boushey, *State of Working America*,

2002–2003, 286–91. On gains from the stock market, see Wolff, "Recent Trends in Wealth Ownership, 1983–1998."

6. Assessing the importance of different factors in causing increased economic inequality since the 1970s is a complex task that has preoccupied numerous scholars. Many debates are ongoing. For a good overview of the causes of inequality debate, see Richard Freeman, *When Earnings Diverge: Causes, Consequences, and Cures for the New Inequality in the U.S.*, National Policy Association, 1997. On the subject of taxes and increased inequality, see David Cay Johnston, *Perfectly Legal: The Secret Campaign to Rig Our Tax System to Benefit the Super Rich—and Cheat Everybody Else* (New York: Portfolio 2003). For a good overview of the role of public policy in increasing inequality, see Gabriel Lenz, "The Policy-Related Causes and Consequences of Inequality," unpublished paper, January 2003. On the role of corporate restructuring and deregulation in increasing inequality, see Leslie McCall, "Corporate Restructuring and Rising Inequality: What is the Connection and Does It Matter?" *Demos*, December 2003.
7. Dinesh D'Souza, "The Virtue of Prosperity," *Hoover Digest*, no. 1 (2001). For an excellent analysis of the conservative defense of economic inequality, and a critique of that defense, see Shelly Arseneault and Donald Matthewson, "Conservative Dilemmas: Problems with the Ideological Defense of Income Inequality," unpublished paper, August 2003.
8. For an overview of recent research on this topic, see Alan Krueger, "The Apple Falls Close to the Tree, Even in the Land of Opportunity," *New York Times*, 14 November 2002, C2.
9. Much of the research exploring the possible social consequences of inequality has been supported by the Russell Sage Foundation. Most of it remains in the form of unpublished papers (as of this writing). See Larry Bartels, "Partisan Politics and U.S. Income Distribution," unpublished paper, May 2003; Theda Skocpol, "Voice and Inequality: The Contemporary Transformation of American Democracy," unpublished paper, undated; Jonathan Schwabish and Timothy Smeeding, "Income Distribution and Social Expenditures: A Crossnational Perspective," unpublished paper, May 2003; and Bruce Western and Becky Pettit, "Black-White Wage Inequality, Employment Rates, and Incarceration," unpublished paper, December 2002.
10. Harvey Araton, "After Perfection at the Age of 12, What Next?" *New York*

Times, 26 August 2001, 8; Richard Sandomir, "Hey, ABC, It Was Not a One-Team Tournament," *New York Times,* 27 August 2001, D2; and Robert D. McFadden, "Star Is 14, So Bronx Team Is Disqualified," *York Times,* 1 September 2001, A1.

11. Ibid.
12. Mark Hyman, "Little League, Big Dollars," *Business Week,* 4 September 2000, 72–73; Patrick McGeehan, "Wall Street's Spiritual Lift from the Boys of Summer," *New York Times,* 26 August 2001, 2; "Bronx Team Hits an ESPN High," *New York Times,* 25 August 2001, 2; and Joseph P. Fried, "Life after Little League for a Tarnished Star," *New York Times,* 15 December 2002, 55.
13. Don Weiskopf, "How Barry Bonds Emerged as Baseball's Premier Slugger," *Baseball Play America,* August 5, 1999; and Charles Yesalis quoted in Dayn Perry, "Pumped-up Hysteria," *Reason* 8, no. 34 (January 2003): 32.
14. Tom Verducci, Don Yaeger, et al., "Totally Juiced," *Sports Illustrated,* 3 June 2002, 34–45.
15. Rick Reilly, "The 'Roid to Ruin," *Sports Illustrated,* 21 August 2000, 92.
16. Verducci, Yaeger, et al., "Totally Juiced."
17. Ibid.
18. Ibid.
19. Ibid.; and Reilly, "The 'Roid to Ruin."
20. "Superhuman Heroes," *Economist,* 6 June 1998, 10–12.
21. Douglas Looney, "Cycling's Premier Race: Tour de Drugs?" *Christian Science Monitor,* 24 July 1998, B8; and E. M. Swift, "Drug Pedaling," *Sports Illustrated,* 5 July 1999, 60.
22. Ibid.
23. Rick Reilly, "Lance Armstrong: Sportsman of the Year," *Sports Illustrated,* 16 December 2002, 52. This figure is for 2001.
24. Robert H. Frank and Philip J. Cook, *The Winner-Take-All Society: Why the Few at the Top Get So Much More Than the Rest of Us* (New York: Penguin Books, 1996), 37.
25. Maureen Tkacik, "Many at Post Office Reject Spin on Cycling," *Wall Street Journal,* 29 July 2002, B1.
26. Ibid.
27. The survey was done by the Good Work Project at Harvard University. See Howard Gardner, Mihaly Csikszentmihalyi, and William Damon,

Good Work: When Excellence and Ethics Meet (New York: Perseus, 2001), 128. In the Pew survey, however, only 10 percent of journalists cited a lack of ethics or accurate/factual reporting as the biggest problem confronting journalism. Over three quarters said that there were ongoing efforts to address ethics in their news organizations. See "*Striking the Balance, Audience Interests, Business Pressures, and Journalists' Values*," Pew Center for the People and the Press, 30 March 1998.

28. Howard Kurtz, "Stranger Than Fiction," *Washington Post*, 13 May 1998, A1. See also Josh Getlin, "New Life for Adage: Never Let Facts Get in Way of a Good Story," *Los Angeles Times*, 30 June, 1998, A5; and William Powers, "The Use and Abuse of Fiction," *National Journal*, 27 June, 1998, 1520.
29. Kurtz, "Stranger Than Fiction," A1.
30. For a survey of trends in journalists' salaries, see Anne Colamosca, "Pay for Journalists," *Columbia Journalism Review* (July/August 1999): 24.
31. Other cases in the 1990s include: Ken Hamblin, Julie Amparano, Michael Kinney, Karen Mamome, Antonietta Palleschi, Albert Flores, Edwin Chen, Fox Butterfield, Bob Morris, Gregory Freeman, Laura Parker, and Kim Stacy. On the cases from the mid and early 1990s, see Trudy Lieberman, "Plagiarize, Plagiarize, Plagiarize . . ." *Columbia Journalism Review* (July/August 1995): 21.
32. Robert Frank, *Luxury Fever: Why Money Fails to Satisfy in an Age of Excess* (New York: The Free Press, 1999), 129.
33. On the link between hierarchy and happiness, see Gordon D. A. Brown, Jonathan Gardner, Andrew Oswald, and Jing Qian, "Rank Dependence in Pay Satisfaction," June 3, 2003 (unpublished paper). On the link between stress and hierarchy, see Robert Spolasky, *Why Zebras Don't Get Ulcers: An Updated Guide to Stress, Stress-Related Diseases, and Coping* (New York: Henry Holt, 1998); on the link between health and inequality, see H. Bosma, M. G. Marmot, H. Hemingway, A. G. Nicholson, E. Brunner, A. Stansfeld, "Low Job Control and Risk of Coronary Heart Disease in Whitehall II (Prospective Cohort) Study," *British Medical Journal* 314 (1997): 558–65. (For an overview of the continuing research related to the Whitehall study, see *www.ucl.ac.uk/epidemiology/white/white.html*) See also James Lardner's article about the pioneering work of British health researcher Richard Wilkinson. James Lardner, "Inequality Meets Epidemiology,"

Inequality.org (at *www.inequality.org*). Debates on health and inequality are complex and contested. For a good bibliography of existing research, see the John D. and Catherine T. MacArthur Research Network on Socioeconomic Status and Health. (*www.macses.ucsf.edu*).

34. Frank, *Luxury Fever,* 145.
35. "Honesty/Ethics in Professions," The Gallup Organization, 1997.
36. General Social Survey, The National Opinion Research Center; and Jeffrey Jones, "Americans Express Little Trust in CEOs of Large Corporations or Stockbrokers," Gallup New Service, 17 July 2002. For analysis and discussion of GSS data on trust, see Uslaner, *The Moral Foundations of Trust.*
37. Uslaner, *The Moral Foundations of Trust,* 33.
38. Everett Carll Ladd and Karlyn H. Bowman, *Attitudes Toward Economic Inequality* (Washington, D.C.: AEI Press, 1998), 95 and 60–73.
39. Claudia Goldin and Robert Margo, "The Great Compression: The Wage Structure in the United States at Mid-Century," National Bureau of Economic Research, August 1991; and Paul Krugman, "For Richer," *New York Times Magazine,* 22 October 2002, 62.
40. David Callahan, *Kindred Spirits: Harvard's Extraordinary Business School Class of 1949 and How They Transformed American Business* (New York: John Wiley & Sons, 2002).
41. Uslaner, *The Moral Foundations of Trust,* 181.
42. For evidence that residential segregation has increased along with inequality, see Susan Mayer, "The Effect of Geographic Distribution of Income Inequality on Children's Educational Attainment," *Northwestern University/University of Chicago Joint Center for Poverty Research,* 2001.
43. Edward Blakely and Mary Gail Snyder, *Fortress America: Gated Communities in the United States* (Washington, D.C.: The Brookings Institution, 1997). See also Haya El Nasser, "Gated Communities More Popular, and Not Just for the Rich," *USA Today,* 15 December 2002, A1.

Chapter Four: A Question of Character

1. Stephanie Mehta, "Birds of a Feather," *Fortune,* 14 October 2002, 197.
2. On Sullivan's rise, see "Scott Sullivan: Master of the Merger," *Oswego Magazine* (spring 1999): 2; Lynn Jeter, *Disconnected: Deceit and Betrayal at WorldCom* (New York: John Wiley & Sons, 2003), 54; and Beatrice Garcia, "A Fast Rise, Faster Fall," *SiliconValley.com,* 27 June 2002.

3. For the most thorough and authoritative investigation of the WorldCom scandal, see United States Bankruptcy Court, Southern District of New York, "First Interim Report of Dick Thornburgh, Bankruptcy Examiner," 4 November 2002. See also Julie Homer, "How Did We Get Here?," *CFO* (October 2002): 40. For a concise dissection of WorldCom's fraud, see Frank Partnoy, *Infectious Greed: How Deceit and Risk Corrupted the Financial Markets* (New York: Times Books, 2003), 366–73.
4. Stephen L. Carter, *Integrity* (New York: HarperPerennial, 1997), 7.
5. There is a large (and contentious) literature that examines the nature of moral development and the role of different institutions and factors in shaping human character. A good deal of this literature revolves around the work of Kohlberg, who conducted empirical studies on moral development, identifying six stages through which such development occurs as people mature into adults. Other parts of this literature examine the particular influences of school, family, peers, and culture. There is also a literature on social norms and society, as well as a growing body of work that argues that biology is a critical determinant of human morality. See, for example, F. Clark Power, Ann Higgins, and Lawrence Kohlberg, *Lawrence Kohlberg's Approach to Moral Education* (New York: Columbia University Press, 1989); Elliot Turiel, *The Culture of Morality: Social Development, Context, and Conflict* (New York: Cambridge University Press, 2002); Thomas Lickona, *Educating for Character: How Our Schools Can Teach Respect and Responsibility* (New York: Bantam Books, 1991); Larry Nucci, ed., *Moral Development and Character Education: A Dialogue* (Berkeley: McCutchan, 1989); Melanie Killen and Daniel Hart, eds., *Morality in Everyday Life: Developmental Perspectives* (New York: Cambridge University Press, 1995); Leonard Katz, *Evolutionary Origins of Morality: Cross-Disciplinary Perspectives* (Thorverton, UK: Imprint Academic, 2000); and Amitai Etzioni, "Social Norms: Internalization, Persuasion, and History," *Law and Society Review* 34, no. 1 (2000): 157–78.
6. There is much survey data and secondary literature on American values. For good overviews of public opinion on key values issues, see "Issues in the 2000 Election: Values," The Washington Post/Kaiser Family Foundation/Harvard University, September 2000; "Pew Values Update: American Social Beliefs 1997–1987," The Pew Research Center for the People and the Press, 20 April 1998; and "American Values: 1998 National Survey of

Americans on Values," Washington Post/Kaiser/Harvard Survey Project. See also the General Social Survey and the World Values Survey. In addition, see The Post-Modernity Project, *The State of Disunion: 1996 Survey of American Political Culture* (Ivy, Va.: In Medias Res Foundation, University of Virginia, 1996); Daniel Yankelovich, "How Changes in the Economy Are Reshaping American Values," in Henry J. Aaron, Thomas E. Mann, and Timothy Taylor, eds., *Values and Public Policy* (Washington, D.C.: The Brookings Institution, 1994); and Alan Wolfe, *One Nation After All: What Middle-Class Americans Really Think About* (New York: Penguin Books, 1998).

7. John Leo, "One Tin Slogan," *US News & World Report*, 22 January 2001, 13.
8. Elizabeth Michaelson, "Director Andrew Douglas Retools U.S. Army's Image," *SHOOT*, 23 February 2001, 7.
9. Leo, "One Tin Slogan," 7.
10. For more on individualism in the U.S., see Robert Bellah, *Habits of the Heart: Individualism and Commitment in American Life* (Berkeley: University of California Press, 1996); Barry Alan Shain, *The Myth of American Individualism* (Princeton: Princeton University Press, 1994); and Robert Bork, *Slouching Toward Gomorrah: Modern Liberalism and American Decline* (New York: The Free Press, 1996).
11. See Ronald Inglehart, *Culture Shift in Advanced Industrial Society* (Princeton: Princeton University Press, 1989); Inglehart, *Modernization and Postmodernization: Cultural, Economic, and Political Change in 43 Societies* (Princeton: Princeton University Press, 1997); and Daniel Yankelovich, "The Shifting Direction of America's Cultural Values," address to DYG's annual SCAN Conference, New York City, 29 May 1998. For another treatment of individualism's global reach, see Thomas M. Franck, *The Empowered Self: Law and Society in the Age of Individualism* (New York: Oxford University Press, 2000).
12. "The Widening Rift between Corporations and Society," interview with James Maxmin and Shoshana Zuboff, *Working Knowledge*, 14 October 2002. See their discussion of individualism in Shoshana Zuboff and James Maxmin, *The Support Economy: Why Corporations Are Failing Individuals and the Next Episode of Capitalism* (New York: Viking, 2002), 93–117.
13. Christopher Lasch, *The Culture of Narcissism: American Life in an Age of Diminishing Expectations* (New York: W. W. Norton & Company, 1979). For

an overview of the rise of the Christian right, see Steve Bruce, *The Rise and Fall of the New Christian Right: Conservative Protestant Politics in America 1978–1988* (New York: Oxford University Press, 1990).

14. David Brooks, *Bobos in Paradise: The New Upper Class and How They Got There* (New York: Simon & Schuster, 2000).
15. Poll cited in Hendrik Hertzberg, "The Short Happy Life of the American Yuppie," in Nicolaus Mills, ed., *Culture in an Age of Money: The Legacy of the 1980s in America* (Chicago: Ivan R. Dee, 1990), 71.
16. Ibid., 82.
17. The survey data in this area across different polls paints a somewhat murky picture. For example, a 1990 Roper poll indicated a doubling of the percent of Americans between 1975 and 1990 who said that being wealthy came closest to expressing their "personal idea of success," but this goal ranked far behind other goals, such as being true to God or being a good parent. See Everett Carll Ladd and Karlyn H. Bowman, *Attitudes Toward Economic Inequality* (Washington, D.C.: AEI Press, 1998), 51.
18. Alexander W. Astin et al., *The American Freshman: Thirty-Year Trends* (Los Angeles: Higher Education Research Institute, 1997), 46–47.
19. Michael Lewis, *Liar's Poker* (New York: Penguin Books, 1989), 24 and 9.
20. Michael Lewis, "Jonathan Lebed's Extracurricular Activities," *New York Times Magazine*, 25 February 2001, 26.
21. Alissa Quart, *Branded: The Buying and Selling of Teenagers* (New York: Perseus, 2003). Quart quoted in William Holstein, "Marketers Crank It Up for a New Generation," *New York Times*, 26 January 2003, III6.
22. Juliet Schor, *The Overspent American: Upscaling, Downshifting, and the New Consumer* (New York: Basic Books, 1998).
23. On estimates of income needed, see data from the Roper Center's iPOLL online public opinion database. Survey about aspirations cited in Schor, *The Overspent American*, 13. On reports of insufficient income, see Ladd and Bowman, *Attitudes Toward Economic Inequality*, 95. As for gaps between aspiration and income in previous eras, note that one scholar estimated over sixty years ago that Americans at every income level wanted to be making 25 percent more money. H. F. Clark, as cited in Robert Merton, *Social Theory and Social Structure* (New York: The Free Press, 1957), 136.
24. On working longer hours, see Lawrence Mishel, Jared Bernstein, and Heather Boushey, *State of Working America, 2002–2003* (Washington, D.C.:

Economic Policy Institute, 2002), 115; and Juliet Schor, *The Overworked American: The Unexpected Decline of Leisure.* On credit card debt, see Tamara Draut and Javier Silva, *Borrowing to Make Ends Meet* (New York: Demos, 2003).

25. Much of the following account, including those individuals quoted, draws from the excellent article on Silverman's life and death by Beth Landman Kell, "The Man Who Had Everything," *New York,* 2 December 2002.
26. For the classic account, see Richard Hofstadter, *Social Darwinism in American Thought* (Boston: Beacon Press, 1992). See also Carl N. Dengler, *In Search of Human Nature: The Decline and Revival of Darwinism in American Social Thought* (New York: Oxford University Press, 1992).
27. On working hard and admiration for the rich, see findings from a variety of surveys in Ladd and Bowman, *Attitudes Toward Economic Inequality,* 53–57; on blame for failure, see Tamara Draut, *New Opportunities: Public Opinion on Poverty, Income Inequality, and Public Policy, 1996–2002* (New York: Demos, 2002), 7. Other surveys on the causes of poverty report similar findings, although these findings are not consistent across all polling. See, for example, surveys reported by Ladd and Bowman, *Attitudes Toward Economic Inequality,* 52. Daniel Yankelovich's data also underscores what he says is a "trend toward Social Darwinism." Yankelovich comments that "unequal results are no longer deemed to be society's fault." See Daniel Yankelovich, "How American Individualism Is Evolving," *Public Perspective,* February/March 1998.
28. Marvin Olasky, *The Tragedy of American Compassion* (Washington, D.C.: Regnery Publishing, 1992).
29. For one of the best discussions of the American Dream ideology, see Jennifer L. Hochschild, *Facing Up to the American Dream: Race, Class, and the Soul of the Nation* (Princeton: Princeton University Press, 1995). See also Hochschild's earlier book, *What's Fair: American Beliefs About Distributive Justice* (Cambridge, Mass.: Harvard University Press, 1981). For an excellent overview of American exceptionalism, see Seymour Martin Lipset, *American Exceptionalism: A Double-Edged Sword* (New York: W. W. Norton, 1997).
30. Merton, *Social Structure and Social Theory,* 142.
31. Mike Maffei, "Enron Employees Share Blame," *Crusader,* 1 February 2002.
32. Partnoy, *Infectious Greed,* 297.

33. Brian Cruver, *Anatomy of Greed: The Unshredded Truth from an Enron Insider* (New York: John Wiley & Sons, 2002), 24. See also Mimi Swartz with Sherron Watkins, *Power Failure: The Inside Story of the Collapse of Enron* (New York: Doubleday, 2003); Robert Bryce and Molly Ivins, *Pipe Dreams: Greed, Ego, and the Death of Enron* (New York: Public Affairs, 2002); Loren Fox, *Enron: The Rise and Fall* (New York: John Wiley, 2002); and Peter C. Fusaro and Ross M. Miller, *What Went Wrong at Enron: Everyone's Guide to the Largest Bankruptcy in U.S. History* (New York: John Wiley, 2002).
34. Mike Tolson and Alan Bernstein, "Power Failure," *Houston Chronicle*, 10 February 2002, 1.
35. Carol Hazard, "Forced Rankings in Legal Tangle," *Richmond-Times Dispatch*, 4 June 2002, A1; and Geoffrey Colvin, "We Can't All Be Above Average," *Fortune*, 13 August 2001, 48.
36. Cruver, *Anatomy of Greed*, 64. See also L. M. Sixel, "Enron Rating Setup Irks Many Workers," *Houston Chronicle*, 26 January 2001, 1.
37. David Streitfeld and Lee Romney, "Enron's Run Tripped by Arrogance, Greed," *Los Angeles Times*, 27 January 2002, 1; and John Byrne, "At Enron, 'The Environment Was Ripe for Abuse,'" *Businessweek Online*, 15 February 2002.
38. Cruver, *Anatomy of Greed*, 69.

Chapter Five: Temptation Nation

1. For an example of one of William Bennett's rare columns that dealt with the corporate scandals, see "Capitalism and Moral Education," *Chicago Tribune*, 28 July 2002, C9. In addition to Bennett's writing on values, an excellent exposition of the conservative argument on values can be found in Ben Wattenberg's writing. See Ben Wattenberg, *Values Matter Most: How Democrats, or Republicans, or a Third Party, Can Win and Renew the American Way of Life* (New York: Free Press, 1995). Wattenberg has stated his position in these blunt terms: "I believe that the values situation in America has worsened. I believe that government has played a big role in allowing values to erode. I believe that in the governmental arena, most of the blame for what has happened goes to liberal guilt peddlers whose remedies almost invariably involved what has been called 'something for nothing.' This erosion is not irreversible. I believe that what government has caused, government can cure. I believe that what liberalism has caused, conservatism can cure.

I believe that liberalism, which has contributed so much to America, might still change, and help lead."

2. Robert Bork, *Slouching Toward Gomorrah: Modern Liberalism and American Decline* (New York: The Free Press, 1996), 8.
3. David Staples, "A Telecom Prophet's Fall from Grace," *Edmonton Journal,* 28 July 2002, 136.
4. Devin Leonard, "The Adelphia Story," *Fortune,* 12 August 2002, 136.
5. Landon Thomas, Jr., "Former Chairman of Qwest Agrees to Give Up Gains from Stock Deals," *New York Times,* 14 May 2003, C1.
6. U.S. Securities and Exchange Commission, *SEC* v. *KPMG LLP,* 29 January 2003.
7. Ibid. My account of the events at KPMG relating to Xerox is drawn entirely from the SEC's complaint.
8. U.S. Securities and Exchange Commission, "In the Matter of KPMG Peat Marwick LLP," 19 January 2001.
9. "KPMG Statement Regarding SEC Action in Xerox Matter," 29 January 2003.
10. Gary John Previts and Barbara Dubis Merino, *A History of Accounting in the United States: The Cultural Significance of Accounting* (Columbus: Ohio State University Press, 1998), 353–54.
11. On the history of the accounting industry and its various scandals see, for example, Mike Brewster, *Unaccountable: How the Accounting Profession Forfeited a Public Trust* (New York: John Wiley & Sons, 2003); David Grayson Allen and Kathleen McDermott, *Accounting for Success: A History of Price Waterhouse in America, 1890–1990* (Cambridge, Mass.: Harvard Business School Press, 1993); and Philip B. Chenok, *Foundations for the Future: The AICPA from 1980 to 1995* (Stamford, Conn.: JAI Press, Inc., 2000).
12. Richard L. Hudson, "The Deregulator," *Wall Street Journal,* 12 January 1984, 1.
13. Ibid.; and Kenneth B. Noble, "Behind the Dispute over the SEC," *New York Times,* 21 April 1982, D13.
14. Dean Foust, "The Big Six Are in Big Trouble," *Business Week,* 6 April 1992, 78; and Thomas McCarroll, "Who's Counting," *Time,* 13 April 1992, 48.
15. Lynn Turner, "Independence: A Covenant for the Ages," remarks to International Organization of Securities Commissions, Stockholm, Sweden, 28 June 2001.

16. Cited in Frank Partnoy, *Infectious Greed: How Deceit and Risk Corrupted the Financial Markets* (New York: Times Books, 2003), 206.
17. Chenok, *Foundations for the Future,* 8; and Jeffrey Marshall, "CPAs as Consultants: Conflict of Interest?" *United States Banker* (November 1991): 24.
18. Marshall, "CPAs as Consultants," 24. See also Mark Stevens, *The Big Six: The Selling Out of America's Top Accounting Firms* (New York: Simon & Schuster, 1991).
19. U.S. Securities and Exchange Commission, "Revision of the Commission's Auditor Independence Requirements," *Federal Securities Law Reports,* 13 December 2000, 84,005.
20. "Accounting Industry: Total Contributions per Election Cycle, 1989–2001"; 30 January 2002; and "Accounting Industry: Lobbying Expenditures, 1997–2001," *Open Secrets,* Center for Responsive Politics, 30 Janaury 2002.
21. "AICPA Establishes Mandatory Quality Control System," *Business Wire,* 1 February 2000.
22. Quoted in Jane Mayer, "The Accountant's War," *New Yorker,* 22 April 2002, 64.
23. For a good discussion of why Spitzer was able to step into the breach, see Partnoy, *Infectious Greed,* 287.
24. U.S. Congress, House Committee on Ways and Means, Subcommittee on Oversight, *The "Tax Gap" and Taxpayer Noncompliance* (Washington, D.C.: Government Printing Office, 1990), 4; Ralph Vartabedian, "Unpaid Tax Total Put at $195 Billion a Year by IRS," *Los Angeles Times,* 2 May 1998, 1; Kathy M. Kristof, "IRS to Increase Number of Audits," *Los Angeles Times,* 1 March 2002, C1; and David R. Francis, "Now, Fewer Tax Cheats in IRS Net," *Christian Science Monitor,* 12 April 2002, 1.
25. Amy Feldman and Joan Caplin, "Should You Cheat on Your Taxes?" *Money* (April 2001): 108–13; and Lori Nitschke, "Congress Still Skeptical of 'Kinder, Gentler' IRS," *CQ Weekly* 57, 8 May 1999, 1065–69.
26. Charles Lewis and Bill Allison, *The Cheating of America* (New York: HarperCollins Perennial, 2002), 261. This book provides an excellent overview of the problem of growing tax evasion in the United States. For other overviews, see David Cay Johnston, *Perfectly Legal: The Secret Campaign to Rig Our Tax System to Benefit the Super Rich—and Cheat Everybody Else* (New York: Portfolio 2003); and Donald Bartlett and James Steele, *The*

Great American Tax Dodge: How Spiraling Fraud and Avoidance Are Killing Fairness, Destroying the Income Tax, and Costing You (New York: Little Brown, 2000). See also Jonathan Weisman, "IRS Eases Up on Tax Cheats," *USA Today*, 9 April 2001, A1; and Janet Novack, "Are You a Chump?" *Forbes* 167, 5 March 2001, 122–29.

27. Tim Dickenson, "Taxfree Inc.," *Mother Jones*, March/April 2000, 18–19; Howard Gleckman et al., "Tax Dodging: Enron Isn't Alone," *Business Week*, March 4, 2002, 40–41; and John D. McKinnon, "New Penalties, Constraints Seen for Tax Dodgers," *Wall Street Journal*, 14 March 2002, A2.
28. David Cay Johnston, "Tax Inquiries Fall as Cheating Increases," *New York Times*, 14 April 2003, A1; David Cay Johnston, "Departing Chief Says the IRS Is Losing Its War on Tax Cheats," *New York Times*, 5 November 2002, A1; and David Cay Johnston, "Affluent Avoid Scrutiny on Taxes Even as IRS Warns of Cheating," *New York Times*, 7 April 2002, A1.
29. Lewis and Allison, *The Cheating of America*, 14; and Johnston, "Affluent Avoid Scrutiny on Taxes," A1. Survey cited in Feldman and Caplin, "Should You Cheat on Your Taxes?" 108–13.
30. Lewis and Allison, *The Cheating of America*, 260.
31. Mark C. Winings, "Ignorance Is Bliss, Especially for the Tax Evader," *Journal of Criminal Law and Criminality* 84 (1993): 575.
32. David R. Francis, "IRS Swings Back into Old Role: Enforcer," *Christian Science Monitor*, 16 September 2002, 17.
33. David R. Francis, "Tax Counselor vs. Tax Cop," *Christian Science Monitor* 91, 2 June 1999, 8; and Mike McNamee, "Throwing Out the Taxes with the Taxman," *Business Week*, 22 June 1998, 190.
34. Johnston, "Departing Chief Says the IRS Is Losing Its War on Tax Cheats."
35. For the best discussion of this, see Sidney Verba, Henry E. Brady, and Kay L. Schlozman, *Voice and Equality: Civic Voluntarism in American Politics* (Cambridge, Mass.: Harvard University Press, 1995). See also Steven J. Rosenstone and John M. Hansen, *Mobilization, Participation, and Democracy in America* (Boston: Prentice Hall, 1993).
36. On spending by conservative think tanks, see Andrew Rich, "Think Tanks and the Politics of Policy Analysis," unpublished paper, 31 August 2002. Rich's data is for 1996. The amount is surely greater today. For more

on conservative think tanks, see David Callahan, *$1 Billion for Ideas: Conservative Think Tanks in the 1990s* (Washington, D.C.: National Committee for Responsive Philanthropy, March 1999). For a historical perspective on the uses and misuses of social science, see James Allen Smith, *The Idea Brokers: Think Tanks and the Rise of the New Policy Elite* (New York: The Free Press, 1990).

37. Shawn Zeller, "'Free Market' Crusaders," *National Journal*, 11 July 2003.
38. *Influence, Inc.: The Bottom Line on Washington Lobbying* (Washington, D.C.: Center for Responsive Politics, 2000); and The Institute for Money in State Politics, *2000 State Election Cycle*, 4 August 2002.

Chapter Six: Trickle-down Corruption

1. The article, by James Wallerstein and Clement J. Wylie, is cited in Robert Merton, *Social Structure and Social Theory* (New York: Free Press, 1957), 144.
2. Regarding people's views on whether their voice matters in politics, see data from the National Election Studies and General Social Survey. There is also a considerable political-science literature that documents and discusses political efficacy. See, for example, Sidney Verba, Henry E. Brady, and Kay L. Schlozman, *Voice and Equality: Civic Voluntarism in American Politics* (Cambridge, Mass.: Harvard University Press, 1995); and Steven J. Rosenstone and John M. Hansen, *Mobilization, Participation, and Democracy in America* (Boston: Prentice Hall, 1993). On public perceptions of the excessive power of corporations in American society, see for example, "Pew Values Update: American Social Beliefs, 1997," The Pew Research Center for the People and the Press, 20 April 1998. On views about fairness of taxes, see National Public Radio/Kaiser Family Foundation/Kennedy School of Government, "National Survey of Americans' Views on Taxes," 2003. On the ability of hard work to get people ahead and for who's underpaid, see, for example, Everett Carll Ladd and Karlyn H. Bowman, *Attitudes Toward Economic Inequality* (Washington, D.C.: AEI Press, 1998), 56 and 20–21.
3. On expectations of wealth, see Ladd and Bowman, *Attitudes Toward Economic Inequality*, 15 and 69–70. On haves and have-nots, see Tamara Draut, *New Opportunities: Public Opinion on Poverty, Income Inequality, and Public*

Policy, 1996–2002 (New York: Demos, 2002), 9–10. That so many people would see themselves among the have-nots is unsurprising, since many Americans reported not having enough money. See also Ladd and Bowman, *Attitudes Toward Economic Inequality*, 95.

4. On fear of poverty, see Draut, *New Opportunities*, 10. On meeting basic needs, see Heather Boushey, Chauna Brocht, Bethney Gundersen, and Jared Bernstein, *Hardships in America: The Real Story of Working Families* (Washington, D.C.: Economic Policy Institute, 2002). On feeling pinched, see Ladd and Bowman, *Attitudes Toward Economic Inequality*, 92. Also, according a survey by the Pew Research Center, 40 percent of Americans agreed in 1997 with the statement: "I often don't have enough money to make ends meet." At the height of the '80s boom, in May 1987, 43 percent of Americans agreed with this statement. See "Pew Values Update: American Social Beliefs, 1997–1987," The Pew Research Center for the People and the Press, 20 April 1998.
5. Neil Fligstein and Taek-Jin Shin, "The Shareholder Value Society: A Review of the Changes in the Working Conditions and Inequality in the U.S., 1976–2000," unpublished paper.
6. David Brooks, "The Triumph of Hope," *New York Times*, 12 January 2003.
7. On anxiety, see Robert Putnam's analysis of DDB Needham Life Style Survey data, *Bowling Alone: The Collapse and Revival of American Community* (New York: Simon & Schuster, 2000), 475. On job satisfaction, see Fligstein and Shin, "The Shareholder Value Society." Evidence of growing insecurity and anxiety is by no means ironclad and this remains a disputed point among scholars. See, for example, Kenneth Deavers, "Downsizing, Job Insecurity, and Wages: No Connection," Employment Policy Foundation, May 1998.
8. Michael Hout, "Money and Morale: What Growing Economic Inequality Is Doing to Americans' View of Themselves and Others," working paper, Survey Research Center, 3 January 2003.
9. Merton, *Social Theory and Social Structure* (New York: The Free Press, 1957), 136–47, 169.
10. Elliot Turiel, *The Culture of Morality: Social Development, Context, and Conflict* (Cambridge, U.K.: Cambridge University Press, 2002), 261 and 266. For

another analysis along somewhat similar lines, see James C. Scott, *Domination and the Arts of Resistance: Hidden Transcripts* (New Haven: Yale University Press, 1990). Scott's book deals extensively with what he calls the "veiled cultural resistance of subordinate groups" and the "infrapolitics of the powerless."

11. Tom R. Tyler, *Why People Obey the Law* (New Haven: Yale University Press, 1990).
12. The interplay between social norms, law, economics, societal values, and compliance with rules has been examined from a variety of angles. One critical—and obvious—observation is that social norms are the key to enforcing rules, since coercion, punishment, and deterrence can never stop everyone who wants to do wrong. Some scholars like Eric Posner argue that law actually plays only a small role in regulating people's behavior. For an overview of some of this work, see Amitai Etzioni, "Social Norms: Internalization, Persuasion, and History," *Law and Society Review* 34, no. 1 (2000): 157–78. See also Eric A. Posner, *Law and Social Norms* (Cambridge, Mass.: Harvard University Press, 2002); and Michael Hechter and Karl-Dieter Opp, eds., *Social Norms* (New York: Russell Sage Foundation, 2001).
13. The 2001 poll was conducted by *Money* magazine. See Amy Feldman and Joan Caplin, "Should You Cheat on Your Taxes?" *Money* (April 2001): 108–13. According to the General Social Survey, over 80 percent of Americans stated in 1998 that it was wrong to cheat on taxes, a percentage that did not change between 1991 and 1998. A 1999 Roper poll commissioned by the IRS Oversight Board found that 87 percent of Americans said that it was not acceptable to cheat on their taxes at all. This number dropped to 76 percent in a 2001 survey, and other results of the poll also indicated decreasing public commitment to tax compliance. (Post-1998 data is not available from the GSS.) See *IRS Oversight Board Annual Report* (Washington, D.C.: Internal Revenue Service, 2002), 13.
14. On people's perception of cheating by others, see "National Survey of Americans' Views on Taxes," National Public Radio/Kaiser Family Foundation/Kennedy School of Government, 2003. On people's views of the fairness of their taxes, see *Public Opinion on Taxes* (Washington, D.C.: American Enterprise Institute, 2003), 12. Public views about the

fairness of the tax system have not yet changed measurably since the late 1970s.

15. U.S. Congress, House Committee on Ways and Means, Subcommittee on Oversight, *The "Tax Gap" and Taxpayer Noncompliance* (Washington, D.C.: Government Printing Office, 1990), 214 and 215.
16. Paul Roberts, "Democrats Want OECD Plan to Nab Runaway Tax Slaves," *Human Events*, 30 July 2001, 17.
17. Tax compliance has been extensively examined by scholars, particularly by economists. On issues of legitimacy in the tax system, and different strategies to improve compliance with tax laws, see Joel Slemrod, ed., *Why People Pay Taxes: Tax Compliance and Enforcement* (Ann Arbor: University of Michigan, 1991). See also Alan Lewis, *The Psychology of Taxation* (London: Martin Robertson, 1982); and Frank A. Cowell, *Cheating the Government: The Economics of Evasion* (Cambridge, Mass.: MIT Press, 1990). On trust and tax compliance specifically, see Ronald Wintrobe, "Tax Evasion and Trust," unpublished paper, 4 June 2001; and Joel Slemrod, "Trust and Public Finance," National Bureau of Economic Research, September 2002. (This paper also argues that tax cheating goes up as the size of government increases.)
18. Tammy Joyner, "Corporate Crime Not Limited to Bigwigs," *Atlanta Journal-Constitution*, 6 August 2002, A1.
19. Tom Neven, "Employee Theft on the Rise, but Shoplifting Losses Fall," *National Home Center News* 27, 5 November 2001, 17–18; Joshua Kurlantzick, "Those Sticky Fingers," *US News & World Report* 130, 4 June 2001, 44; and Greg Winter, "Employee Larceny Is Bigger and Bolder," *New York Times*, 12 July 2000, C1.
20. Kurlantzick, "Those Sticky Fingers," 44.
21. Quoted in Ibid.
22. On job satisfaction and loyalty, see "Most American Workers Satisfied with Their Job," Gallup Poll, 31 August 2001. On trust of senior managers, see "Declining Trust Threatens Corporate Competitiveness," *HR Daily News*, 30 June 2002. On misconduct in the workplace, see *1999 National Business Ethics Study*, Walker Information Inc./The Hudson Institute, September 1999; and *2000 Organizational Integrity Survey*, KPMG, 2000. See also Jeffrey L. Seglin, "The Ethics Policy: Mind-Set Over Matter," *New York Times*, 16 July 2000, C4; *2000 National Business Ethics Survey* (Washington, D.C.: Ethics Resource Center, 2000). On stress and identity at work, see "Most American

Workers Satisfied With Their Job"; and "Are You Feeling a Lot of Stress at Work?" Poll Analysis, 15 October 2002, Gallup. On declining job satifisfaction and loyalty, see "Special Consumer Survey Report: Job Satisfaction on the Decline," The Conference Board, July 2002.

23. Quoted in David Wessel, "What's Wrong?" *Wall Street Journal,* 20 June 2002.
24. On estimates of music piracy, see D. C. Denison, "The Legacy of Napster Shutdown for Service Looms but File-Sharing Approach Very Much Alive and Thriving," *Boston Globe,* 16 May 2002, E1. On survey findings, see Edna Gundersen, "Piracy Has Its Hooks In," *USA Today,* 6 May 2003, 1D.
25. Nick Wingfield, "Napster: Boy, Interrupted," *Wall Street Journal,* 1 October 2002, B1; Mick Brady, "The End of the Napster Revolution," *E-Commerce Times,* 3 August 2000; and Gundersen, "Piracy Has Its Hooks In," 1D.
26. Jonathan Last, "The Ethics of Napster," *Belief Net* (undated).
27. Gundersen, "Piracy Has Its Hooks In," 1D.
28. Brady, "The End of the Napster Revolution."
29. Stanley A. Miller and Dan Egan, "College Students Bond over File-Swapping Suit," *Milwaukee Journal-Sentinel,* 4 May 2003, 1.
30. Monica Davey, "The Cable Crusader," *Chicago Tribune,* 18 January 1999, 1.
31. Chicago city official quoted in ibid; and Jack Spillane, "City Police Face Cable Theft Probe," *SouthCoast Today,* 22 February 2001.
32. Larry Seben, "Comcast Enlists Kana for Customer Support in Cable-Satellite War," *CRMDaily.com,* 28 August 2001.
33. Davey, "The Cable Crusader," 1.
34. Foundation for Taxpayers and Consumers Rights, "Auto Insurers' Excessive Profits in California Costing Consumers Billions," news release, 21 October 1998.
35. Kenneth Lovett, "D.A.: Road to Insure Scam Leads Upstate," *New York Post,* 5 April 2001, 25.
36. Ruth Gastel, "Insurance Fraud," *Insurance Issues Update,* April 2003.
37. Quoted in Kenneth Reich, "People Defend Lies to Insurer, Survey Finds," *Los Angeles Times,* 14 November 1991, A15.
38. Credit for the felicitous phrase "Brazilianization" belongs to Edward Luttwak, *The Endangered American Dream* (New York: Simon & Schuster,

1994). On corruption in Brazil, see David V. Fleischer, *Corruption in Brazil: Defining, Measuring, and Reducing* (Washington, D.C.: Center for International and Strategic Studies, 2002).

Chapter Seven: Cheating from the Starting Line

1. Andrew Jacobs, "The June of Loren Easton," *New York Times*, 29 June 1997, XIII1.
2. Kay Hymowitz, "The Parent As Career Coach," *American Educator* (spring 2001).
3. "Report Card 2002: The Ethics of American Youth," Josephson Institute of Ethics, October 2002.
4. "A Portrait of a Generation: 25 Years of Teen Behavior and Attitudes," *Who's Who Among American High School Students* (Lake Forest, Ill.: Educational Communications, Inc., 1995), 14; *Who's Who Among American High School Students, 29th Annual Survey of High Achievers* (Lake Forest, Ill.: Educational Communications, Inc., 1998). See also F. Schab, "Schooling Without Learning: Thirty Years of Cheating in High School," *Adolescence* (winter 1991): 841.
5. "Cheating and Succeeding: Record Numbers of Top High School Students Take Ethical Shortcuts," *Who's Who Among American High School Students, 29th Annual Survey of High Achievers*, 1.
6. "A Portrait of a Generation," 6.
7. William Fitzsimmons, Marlyn McGrath Lewis, and Charles Ducey, "Time Out or Burn Out for the Next Generation," Harvard University, Office of Undergraduate Admissions, December 2000.
8. Ibid.
9. Ibid.
10. Robert Worth, "Ivy League Fever," *New York Times*, 24 September 2000, WC1.
11. Pamela Kidder-Ashley, James R. Deni, and Jessica B. Anderton, "Learning Disabilities Eligibility in the 1990s: An Analysis of State Practices," *Education* 121 (fall 2000): 65.
12. Ibid., 70.
13. Jane Gross, "Paying for a Disability Diagnosis to Gain Time on College Boards," *New York Times*, 26 September 2002, A1.
14. Louis Menand, "The Thin Envelope," *New Yorker*, 7 April 2003, 90; and

Ethan Bronner, "College Applicants of '99 Are Facing Stiffest Competition," *New York Times*, 12 June 1999, A1.

15. James Wallace and Jim Erickson, *Hard Drive: Bill Gates and the Making of the Microsoft Empire* (New York: HarperPerennial, 1992), 259–60.
16. Robert Frank and Philip Cook, *The Winner-Take-All Society* (New York: Penguin, 1996), 157.
17. U.S. Census Bureau, *The Big Payoff: Educational Attainment and Synthetic Estimates of Work-Life Earnings* (Washington, D.C.: Government Printing Office, 2002).
18. Greg Winter, "College Loans Rise, Swamping Dreams," *New York Times*, 28 January 2003, A1; and Arthur Levine and Jana Nidiffer, *Beating the Odds: How the Poor Get to College* (San Francisco: Jossey-Bass Publishers, 1996), 35.
19. Frank and Cook, *The Winner-Take-All Society*, 148.
20. Fitzsimmons, Lewis, and Ducey, "Time Out or Burn Out for the Next Generation."
21. "For War on Yale Cheating," *New York Times*, 24 January 1931, 2.
22. Walter W. Ludeman, "A Study of Cheating in Public Schools," *School Board Journal* (March 1938): 45.
23. Charles A. Drake, "Why Students Cheat," *Journal of Higher Education* 12 (November 1941): 419.
24. James A. Blackwell, *On, Brave Old Army Team—West Point, 1951* (Novato, Calif.: Presidio Press, 1996), vi; and Richard C. U'Ren, M.D., *Ivory Fortress: A Psychiatrist Looks at West Point* (Indianapolis: The Bobbs-Merrill Company, Inc., 1974), 90 and 98.
25. William J. Bowers, *Student Dishonesty and Its Control in College* (New York: Columbia University Bureau of Applied Social Research, 1964), 214 and 103.
26. Donald L. McCabe, Linda Klebe Treviño, and Kenneth D. Butterfield, "Cheating in Academic Institutions: A Decade of Research," *Ethics and Behavior* 11 (2001): 220.
27. Quotes and information about the focus groups found in Donald L. McCabe, "Academic Dishonesty Among High School Students," *Adolescence* 34 (winter 1999): 681; and McCabe, "Research Notes for 2002/2003 Assessment Project Study," Center for Academic Integrity, Duke University.
28. See, for example, Valerie A. Wajda-Johnston et al., "Academic Dishonesty at the Graduate Level," *Ethics and Behavior* 11, no. 3 (2001): 287–305; and

Bob S. Brown, "The Academic Ethics of Graduate Business Students: A Survey," *Journal of Education for Business* 70, no. 3: 151.

29. See Sarath Nonis and Cathy Owens Swift, "An Examination of the Relationship between Academic Dishonesty and Workplace Dishonesty: A Multicampus Investigation," *Journal of Education for Business* (November/December 2001): 69–77; R. L. Sims, "The Relationship between Academic Dishonesty and Unethical Business Practices," *Journal of Education for Business* 68, no. 4 (1993): 207–11; and R. A. Fass, "Cheating Plagarism," in W. W. May, ed., *Ethics and Education* (New York: Macmillan, 1990), 170–84.
30. M. Roig and C. Ballew, "Attitudes Toward Cheating of Self and Others by College Students and Professors," *The Psychological Record* 44 (1994): 3–12; and Nonis and Swift, "An Examination of the Relationship between Academic Dishonesty and Workplace Dishonesty," 76.
31. On 1997 study, see "How Candid Are Job Applicants?" *USA Today Magazine* 126 (July 1997): 9. On 2002 study, see Elizabeth Stanton, "If a Resume Lies, Truth Can Loom Large," *New York Times*, 29 December 2002, C8. See also Jeffrey Kluger and Sora Song, "Pumping Up Your Past," *Time*, 10 June 2002, 45. On study by New Jersey firm, see Jeffrey L. Seglin, "Lies Can Have a (Long) Life of Their Own," *New York Times*, 16 June 2002, 3.
32. On HireRight Study, see Joan Fleischer Tamen, "Trust, But Verify," *Sun-Sentinel*, 24 February 2003, 16. On Liar's index, see Ian Parker, "Dishonorable Degrees," *New Yorker*, 4 November 2002, 44–46.
33. Kluger and Song, "Pumping Up Your Past."
34. Rachel Zimmerman, "Bausch Rescinds Its CEO's Bonus," *Wall Street Journal*, 30 October 2002, B2.
35. Gary Smith, "Lying in Wait," *Sports Illustrated*, 8 April 2002, 70–82; and Frank Litsky, "U.S. Olympic Chief Quits Over Her Lies on College Degree," *New York Times*, 25 May 2002, A1.
36. Sandra Baldwin, "They're Ready for Me," *New York Times*, 10 February 2002, 3 and 14.

Chapter Eight: Crime and No Punishment

1. George Lakoff, *Moral Politics: How Liberals and Conservatives Think*, 2d ed. (Chicago: University of Chicago Press, 2002). Quotes from "Left Out by Right Rhetoric," interview with Lakoff, *TomPaine.com*, May 2002.

2. Gary Putka, "Blackboard Jungle: A Cheating Epidemic at a Top High School Teaches Sad Lessons," *Wall Street Journal*, 29 June 1992, A1.
3. David Muha, "Cheating: When Students Cheat," *Rutgers Focus*, 17 March 2000; and Ronald M. Aaron and Robert T. Georgia, "Administrator Perceptions of Student Academic Dishonesty in Collegiate Institutions," *NASPA Journal* 31 (winter 1994): 85.
4. Kevin Bushweller, "Generation of Cheaters," *American School Board Journal*, April 1999. Donald McCabe's research has found similar fears.
5. On the Dwyer case, see Rachel Osterman, "Dartmouth Computer Science Prof Condemns Dean's 'Whitewash,'" *University Wire*, 24 April 2000; Wendy Yu, "Students Relieved by Dartmouth Committee on Standards Verdict," *University Wire*, 27 March 2000; and Benjamin Wallace-Wells, "Research and Its Discontents," *Dartmouth Review*, 13 March 2000.
6. Donald McCabe and Linda Klebe Treviño, "Honesty and Honor Codes," *Academe* 88, no. 1 (January/February 2002): 37; Donald McCabe, "Toward a Culture of Academic Integrity," *Chronicle of Higher Education*, 15 October 1999, B7; Linda Klebe Treviño and Donald McCabe, "Academic Integrity in Honor Code and Non-Honor Code Environments," *Journal of Higher Education* 70 (1999): 211–35.
7. Linda Robertson, "UM Loses a Chance to Teach Players Their Actions Have Consequences," *The MiamiHerald.com*, 31 August 2002.
8. Duncan Mansfield, "UT Professor Says Life Getting Unbearable," Associated Press, 19 September 2000.
9. The financial issues around college athletics are complex. Fans of big-time college athletic programs have long argued that the revenues from televised sports boost the financial standing of the university as a whole. Studies conducted by the NCAA in the 1990s, however, revealed the old truism to be false, with the average Division 1-A football program costing the university about $1 million per year in excess of revenues from ticket sales, TV contracts, and merchandising. See Roger Throw, "Shoe Companies, Tongues Out, Buy Up College Teams Wholesale," *Wall Street Journal*, 17 November 1995, B9. On NCAA revenues, see Allen Burra, "'Amateur' Athletics Are Worth Millions—to NCAA," *Wall Street Journal*, 29 March 1999, A26.
10. James Duderstadt, *Intercollegiate Athletics and the American University* (Ann Arbor: University of Michigan Press, 2000).

11. Ray Johnson, "Inside Longhorn, Inc.: How One College Program Runs the Business," *Fortune*, 20 December 1999, 160.
12. Joe Drape with Ray Glier, "Georgia Suspends Harrick and Withdraws from Postseason," *New York Times*, 11 March 2003, D1.
13. Gallup/CNN/USA Today Poll, June 2003.
14. *2002 Report to the Nation: Occupational Fraud and Abuse* (Austin, Tex.: Association of Certified Fraud Examiners, 2002), 27.
15. Alan Sayre, "Examiners Say Workplace Fraud Is on the Rise," *Marketing News*, 14 September 1998, 22; and Jenni Bergal, "White Collar Crime Costs U.S. Firms Roughly $600 Billion a Year," Knight Ridder/*Tribune Business News*, 5 April 2002.
16. *Lawyer Discipline Report Card* (Washington, D.C.: Lawyer Accountability Project, 2002). See also Megan Barnett, "How to Account for Lawyers," *US News & World Report*, 9 December 2002, 26–27.
17. For more on this subject, see John Caher, "At State Bar Summit, Spitzer Blasts Lawyers on Corporate Scandals," *New York Law Journal*, 23 January 2003; and Anthony E. Davis, "Who Should Regulate Lawyers? Recent Events," *New York Law Journal*, 6 January 2003, 3.
18. Sidney Wolfe, et. al., 2, *125 Questionable Doctors Disciplined by State and Federal Governments: 2000* (Washington, D.C.: 2000), 3.
19. Dagan McDowell, "No Discipline? Why So Few Brokers Get Punished," *TheStreet.com*, 8 June 2000.
20. Kurt Eichenwald, "White-Collar Defense Stance: The Criminal-less Crime," *New York Times*, 3 March 2002, 4.
21. Frank Partnoy, *Infectious Greed: How Deceit and Risk Corrupted the Financial Markets* (New York: Times Books, 2003), 169.
22. Reynolds Holing and William Carlsen, "Hollow Words," *San Francisco Chronicle*, 16 November 1999, 1. For an excellent treatment of the battle against white-collar crime in Silicon Valley, see the *Chronicle*'s five-part investigation of this topic that ran in November 1999, of which this story is one part.
23. Clifton Leaf, "Enough Is Enough," *Fortune*, 18 March 2002, 60.
24. Ibid.
25. Letter from Representative John J. LaFalce, Ranking Democratic Member of House Banking Committee, to President George W. Bush, 31 January 2002.

26. Melody Petersen, "AstraZeneca Pleads Guilty in Cancer Medicine Scheme," *New York Times*, 21 June 2003, C1.
27. Adam Zagorin, "Charlie's an Angel?" *Time*, 3 February 1997, 36–38; and Steven Syre, "Waiting for Corporate Criminals' Time to Fit the Crimes," *Boston Globe*, 1 September 2002, E1.
28. "Wide Disparity in White-Collar Sentences," *USA Today Magazine* (April 2000): 11–12.
29. Kimberly L. Allers, "What to Pack for Prison," *Fortune*, 3 February 2002, 44; Suein L. Hwang, "Facing Doing Time? Just in Case, Consider These Fine Options," *Wall Street Journal*, 21 August 2002, B1; and Russ Mitchell, "White Collar Criminal? Pack Lightly for Prison," *New York Times*, 11 August 2002, C4.
30. Kit R. Roane, "Getting Out of Jail Free," *US News & World Report*, 23 December 2002, 26–28.
31. "Ask the Globe," *Boston Globe*, 24 February 2001, G12.
32. James Kim, "Boesky Gets $50 Million Tax Break," *USA Today*, 25 May 1990, A1; and John D. McKinnon, "Wages of Corporate Sin: Tax Breaks," *Wall Street Journal*, 3 September 2002, A4.
33. Jon Swartz, "Homes of the Rich and Infamous," *USA Today*, 15 July 2002, B3.
34. "White Collar Crime Seldom a Long-Term Black Mark," *Chicago Tribune*, 30 December 1990, 7A.
35. Edward Cohn, "The Resurrection of Michael Milken," *American Prospect*, 13 March 2000, 27–31; and Cynthia Dockrell, "The Insider Milken's Circuitous Route," *Boston Globe*, 17 January 2001, G3.
36. Editorial, "Yes, Pardon Milken," *Wall Street Journal*, 13 December 2000, A26. On Rich, see Tom Juravich and Kate Bronfenbrenner, "Marc Rich Redux," *Nation*, 5 March 2001, 6–7. On government contracts, see Ellen Nakashima, "Study: Contracts Given to Repeat Violators," *Washington Post*, 7 May 2002, A19.
37. "Boy, Did He Get His Comeuppance," *Los Angeles Times*, 10 December 1992, D1; and "Boesky Gets $20 Million in Settlement with Wife," *Chicago Tribune*, 10 June 1993, 10.
38. Susan Antilla, "The Heat's Still on Dennis Levine," *USA Today*, 23 September 1991, B3.

39. "'Bankrupt' Bill Walters Still Lives High on the Hog," *Denver Post*, 18 March 1993, B8.
40. Editorial, "Family Matters," *New York Times*, 10 March 2002, 18; and Bill Husted, "Looker Lauren Bush Has Eyes for Royal Bachelor," *Denver Post*, 8 March 2001, A2.
41. "O'Leary Hired by Vikings," *New York Times*, 12 January 2002, 6.
42. Michael Janofsky, "Harding's Guilty Plea Closes a Saga with No Winners," *New York Times*, 20 March 1994, E2; Jere Longman, "Lines Blur for Kerrigan and Harding," *New York Times*, 6 January 1995, B7; and Alexander Wolff and Christian Stone, "This Week's Sign That the Apocalypse Is upon Us," *Sports Illustrated*, 5 June 1995, 22.

Chapter Nine: Dodging Brazil

1. James Cox, "Inmates Teach MBA Students Ethics from Behind Bars," *USA Today*, 24 May 2001, B1; and Mark Clayton, "For Our Class on White-Collar Crime, a Convict," *Christian Science Monitor*, 3 November 1998, B7. For more on the program, and the view of skeptics, see Benjamin Pimentel, "Jailhouse Seminars," *San Francisco Chronicle*, 17 September 2002, B1.
2. For more on the problem and solutions to the earnings crisis in America, see Holly Sklar, Laryssa Mykyta, and Susan Wefald, *Raise the Floor: Wages and Policies That Work for All of Us* (Boston: South End Press, 2003); and Jared Bernstein and Jeff Chapman, "Time to Repair the Wage Floor," Economic Policy Institute, May 2002. On the number of families who can't make ends meet, see Heather Boushey, Chauna Brocht, Bethney Gundersen, and Jared Bernstein, *Hardships in America: The Real Story of Working Families* (Washington, D.C.: Economic Policy Institute, 2002). On public opinion and the minimum wage, see Tammy Draut, *New Opportunities: Poverty, Income Inequality, and Public Policy* (New York: Demos, 2002). On the impact of the minimum wage, see David Card and Alan B. Krueger, *Myth and Measurement: The New Economics of the Minimum Wage* (Princeton: Princeton University Press, 1995). On the impact and workings of the EITC, see Bruce D. Meyer and Douglas Holtz-Eakin, eds., *Making Work Pay: The Earned Income Tax Credit and Its Impact on Working Families* (New York: The Russell Sage Foundation, 2001.)
3. Jared Bernstein and Dean Baker, *The Benefits of Full Employment: When Markets Work for People* (Washington, D.C.: Economic Policy Institute, 2003).

For more on the link between growth and equity, see Barry Bluestone and Bennett Harrison, *Growing Prosperity: The Battle for Growth with Equity in the 21st Century* (Boston: Houghton Mifflin, 2000).

4. Key background reading on wealth inequities and wealth-building strategies includes Michael Sherraden, *Assets and the Poor: A New American Welfare Policy* (Armonk, N.Y.: ME Sharpe, 1991); Thomas M. Shapiro and Edward N. Wolff, *Assets for the Poor: The Benefits of Spreading Asset Ownership* (New York: The Russell Sage Foundation, 2001); Michelle Miller-Adams, *Owning Up: Poverty, Assets, and the American Dream* (Washington, D.C.: Brookings Institution, 2002); and Bruce Ackerman and Anne Alstott, *The Stakeholder Society* (New Haven, Conn.: Yale University Press, 2000). Ackerman and Alstott's strategy for creating a stakeholder society hinges on giving $80,000 to each American at birth, to be funded through a wealth tax.
5. For a good overview of the idea of public financing, see Ellen S. Miller and Micah L. Sifry, "Democracy That Works: Clean Money Campaign Reform," in Robert Borosage and Roger Hickey, eds., *Next Agenda: Blueprint for a New Progressive Movement* (Boulder, Colo.: Westview Press, 2001). For a concise discussion of free air time, see Paul Taylor and Norman Ornstein, "The Case for Free Air Time," New America Foundation, 1 June 2002.
6. On making elections more competitive, see Stephen Hill, *Fixing Elections: The Failure of America's Winner-Take-All Politics* (New York: Routledge, 2002). For more on election-day registration, see David Callahan and Sarah Tobias, *Expanding the Vote: The Promise and Practice of Election Day Registration* (New York: Demos, 2002).
7. Frank Partnoy, *Infectious Greed: How Deceit and Risk Corrupted Financial Markets* (New York: Times Books, 2003), 394.
8. Richard W. Stevenson and Richard A. Oppel, Jr., "Fed Chief Blames Corporate Greed," *New York Times*, 17 July 2002, A1.
9. See Robert Putnam, *Bowling Alone: The Collapse and Revival of American Community* (New York: Simon & Schuster, 2000); Robert K. Fullinwider, ed., *Civil Society, Democracy, and Civic Renewal* (New York: Rowman & Littlefield, 1999); and Theda Skocpol and Morris P. Fiorina, eds., *Civic Engagement in American Democracy* (Washington, D.C.: Brookings Institution/The Russell Sage Foundation, 1999).
10. This section on sprawl and new urbanism draws heavily from David Callahan and Stephen Heintz, eds., *Quality of Life 2000: The New Politics of*

Work and Community (New York: Demos, 2002), 77–118. Relevant essays in the book include Robert Liberty, "Is the American Dream Endless Sprawl?"; Philip Langdon, "New Development, Traditional Patterns"; "Growth: New Challenges and Opportunities in a New American Landscape—An Interagency Report by the Clinton/Gore Administration"; and Ray Oldenburg, "Our Vanishing 'Third Places.'"

11. Robert Frank, *Luxury Fever: Why Money Fails to Satisfy in an Age of Excess* (New York: The Free Press, 1999), chapters 14–17.
12. Carol Swain, *New White Nationalism in America: Its Challenge to Integration* (New York: Oxford University Press, 2002).
13. Don Oldenberg, "Street-Smart Business Ethics," *Washington Post*, 29 July 1987, D5; and George C. S. Benson, "Codes of Ethics," *Journal of Business Ethics* 8 (1989): 319.
14. Dove Izraeli and Mark S. Schwartz, "What Can We Learn from the U.S. Federal Sentencing Guidelines for Organizational Ethics?" *Journal of Business Ethics* 17 (July 1998): 1045–55; Lou Ann Wolfe, "New Guidelines Put Higher Cost on Breach of Ethics," *Journal Record*, 16 November 1991; Joshua Joseph, *2000 National Business Ethics Survey*, vol. I: *How Employees Perceive Ethics at Work* (Washington, D.C.: Ethics Resource Center, 2000).Daniel E. Palmer and Abe Zakhem, "Bridging the Gap Between Theory and Practice: Using the 1991 Federal Sentencing Guidelines as a Paradigm for Ethics Training," *Journal of Business Ethics* 29 (January 2001): 77–84.
15. Bob Wright, "Restoring Trust," *Vital Speeches of the Day*, 15 December 2002, 150–52.
16. Leah Nathans Spiro, "Ethics and Andersen Didn't Add Up," *Wall Street Journal*, 20 March 2003, D8.
17. *2003 National Business Ethics Survey* (Washington, D.C.: Ethics Resource Center, 2000); *2000 Organizational Integrity Survey*, (New York: KPMG, 2000). See also Jeffrey L. Seglin, "The Ethics Policy: Mind-Set over Matter," *New York Times*, 16 July 2000, C4; and *1999 National Business Ethics Study* (Indianapolis: Walker Information, Inc./The Hudson Institute, September 1999).
18. Robert F. Howe, "Ill Wind Probe: So Far, Less Than Anticipated," *Washington Post*, 1 September 1990, A12.

19. Nancy B. Kurland, "The Defense Industry Initiative: Ethics, Self-Regulation, and Accountability," *Journal of Business Ethics* 12 (February 1993): 137ff. See also Alan P. Mayer-Sommer and Alan Roshwalb, "An Examination of the Relationship between Ethical Behavior, Espoused Ethical Values, and Financial Performance in the U.S. Defense Industry," *Journal of Business Ethics* 15 (December 1996): 1249–74.
20. Lynn Sharp Paine, *Value Shift: Why Companies Must Merge Social and Financial Imperatives to Achieve Superior Performance* (New York: McGraw-Hill, 2003), 24 and 8–23.
21. *Who's Who Among American High School Students, 30th Annual Survey of High Achievers* (Lake Forest, Ill.: Educational Communications, Inc., 2000); and *Who's Who Among American High School Students, 27th Annual Survey of High Achievers* (Lake Forest, Ill.: Educational Communications, Inc., 1996).
22. Ethan Bronner, "Teaching Values Without Taking a Page from the Bible," *New York Times*, 1 June 1999, 17.
23. See, for example, an evaluation of Character Counts in South Dakota, "South Dakota Survey Results, 1998–2000," found on Character Counts Web site at *www.charactercounts.org*. For further evaluation of character education, as well as broader discussion of the theory and practice of character education, see Joan F. Goodman and Howard Lesnick, *The Moral Stake in Education: Contested Premises and Practices* (Boston: Pearson Allyn & Bacon, 2000); Thomas Lickona, *Educating for Character: How Our Schools Can Teach Respect and Responsibility* (New York: Bantam Books, 1991); Larry Nucci, ed., *Moral Development and Character Education: A Dialogue* (Berkeley: McCutchan, 1989); and Madonna Murphy, *Character Education in America's Blue Ribbon Schools* (Lancaster, Pa.: Technomic Publishing Company, 1998).
24. Donald McCabe and Linda Klebe Treviño, "Honesty and Honor Codes," *Academic Online* 88, no. 6 (November/December 2002).
25. "Jail the MBAs?" *Multinational Monitor* 20 (January/February 1999): 4.
26. James Cox, "Inmates Teach MBA Students Ethics from Behind Bars," *USA Today*, 24 May 2001, B1; and Katherine S. Mangan, "MBA's Would Rather Quit Than Fight for Values," *Chronicle of Higher Education*, 20 September 2002, A15.
27. For an inside account, see Amitai Etzioni, "When It Comes to Ethics, B-Schools Get an F," *Washington Post*, 4 August 2002, B4.

28. Paul M. Clikeman and Steven L. Henning, "The Socialization of Undergraduate Accounting Students," *Issues in Accounting Education* 15 (February 2000): 1–17.
29. Serge A. Martinez, "Reforming Medical Ethics Education," *Journal of Law, Medicine, and Ethics* 30 (fall 2002): 452–54.
30. Zachary Karabell, *A Visionary Nation: Four Centuries of American Dreams and What Lies Ahead* (New York, HarperCollins, 2001).

index